VALERIAN

Medicinal and Aromatic Plants – Industrial Profiles

Individual volumes in this series provide both industry and academia with in-depth coverage of one major medicinal or aromatic plant of industrial importance.

Edited by Dr Roland Hardman

Volume 1
Valerian
edited by Peter J. Houghton

Other volumes in preparation

Perilla, Volume 2, edited by He-Ci Yu, Kenichi Kosuna and Megumi Haga

Artemisia, edited by C. Wright
Cannabis, edited by D. Brown
Capsicum, edited by P. Bosland and A. Levy
Cardamom, edited by P.N. Ravindran and K.J. Madusoodanan
Carum, edited by É. Németh
Chamomile, edited by R. Franke and H. Schilcher
Cinnamon and Cassia, edited by P.N. Ravindran and S. Ravindran
Claviceps, edited by V. Křen and L. Cvak
Colchicum, edited by V. Simánek
Curcuma, edited by B.A. Nagasampagi and A.P. Purohit
Eucalyptus, edited by J. Coppen
Evening Primrose, edited by P. Lapinskas
Feverfew, edited by M.I. Berry
Ginkgo, edited by T. van Beek
Ginseng, by W. Court
Illicium and Pimpinella, edited by M. Miró Jodral
Licorice, by L.E. Craker and L. Kapoor
Melaleuca, edited by I. Southwell
Neem, by H.S. Puri
Ocimum, edited by R. Hiltunen and Y. Holt
Piper Nigrum, edited by P.N. Ravindran
Plantago, edited by C. Andary and S. Nishibe
Poppy, edited by J. Bernáth
Saffron, edited by M. Negbi
Stevia, edited by A.D. Kinghorn
Tilia, edited by K.P. Svoboda and J. Collins
Trigonella, edited by G.A. Petropoulos
Urtica, by G. Kavalali

This book is part of a series. The publisher will accept continuation orders which may be cancelled at any time and which provide for automatic billing and shipping of each title in the series upon publication. Please write for details.

VALERIAN
The Genus *Valeriana*

Edited by

Peter J. Houghton

Pharmacognosy Research Laboratories, Department of Pharmacy
King's College London, UK

harwood academic publishers
Australia • Canada • China • France • Germany • India • Japan
Luxembourg • Malaysia • The Netherlands • Russia • Singapore
Switzerland • Thailand • United Kingdom

Amsteldijk 166
1st Floor
1079 LH Amsterdam
The Netherlands

British Library Cataloguing in Publication Data

A catalogue record for this book is available from the British Library

The illustration on the cover is taken from the *Bloemlezing uit het Cruydt-boeck van Rembert Dodoens*, edited by Dr A. Schierbeek, De Hofstad, The Hague, 1941. This is a compilation of *Herbarius oft Cruydt-Boeck* by Rembertus Dodonaeus, Plantijnsche Druckerije van Balthasar Moretus, Antwerp, 1644. We have tried to find the copyright holder of the compilation but were not successful, please contact the publisher in case of any copyright queries.

CONTENTS

PREFACE TO THE SERIES

There is increasing interest in industry, academia and the health sciences in medicinal and aromatic plants. In passing from plant production to the eventual product used by the public, many sciences are involved. This series brings together information which is currently scattered through an ever increasing number of journals. Each volume gives an in-depth look at one plant genus, about which an area specialist has assembled information ranging from the production of the plant to market trends and quality control.

Many industries are involved such as forestry, agriculture, chemical, food, flavour, beverage, pharmaceutical, cosmetic and fragrance. The plant raw materials are roots, rhizomes, bulbs, leaves, stems, barks, wood, flowers, fruits and seeds. These yield gums, resins, essential (volatile) oils, fixed oils, waxes, juices, extracts and spices for medicinal and aromatic purposes. All these commodities are traded world-wide. A dealer's market report for an item may say "Drought in the country of origin has forced up prices".

Natural products do not mean safe products and account of this has to be taken by the above industries, which are subject to regulation. For example, a number of plants which are approved for use in medicine must not be used in cosmetic products.

The assessment of safe to use starts with the harvested plant material which has to comply with an official monograph. This may require absence of, or prescribed limits of, radioactive material, heavy metals, aflatoxins, pesticide residue, as well as the required level of active principle. This analytical control is costly and tends to exclude small batches of plant material. Large scale contracted mechanised cultivation with designated seed or plantlets is now preferable.

Today, plant selection is not only for the yield of active principle, but for the plant's ability to overcome disease, climatic stress and the hazards caused by mankind. Such methods as *in vitro* fertilisation, meristem cultures and somatic embryogenesis are used. The transfer of sections of DNA is giving rise to controversy in the case of some end-uses of the plant material.

Some suppliers of plant raw material are now able to certify that they are supplying organically-farmed medicinal plants, herbs and spices. The Economic Union directive (CVO/EU No 2092/91) details the specifications for the **obligatory** quality controls to be carried out at all stages of production and processing of organic products.

Fascinating plant folklore and ethnopharmacology leads to medicinal potential. Examples are the muscle relaxants based on the arrow poison, curare, from species of *Chondrodendron*, and the antimalarials derived from species of *Cinchona* and *Artemisia*. The methods of detection of pharmacological activity have become increasingly reliable and specific, frequently involving enzymes in bioassays and avoiding the use of laboratory animals. By using bioassay linked fractionation of crude plant juices or extracts, compounds can be specifically targeted which, for example, inhibit blood platelet aggregation, or have antitumour, or antiviral, or any other required activity. With the assistance of robotic devices, all the members of a genus may be readily screened. However, the plant material must be **fully** authenticated by a specialist.

The medicinal traditions of ancient civilisations such as those of China and India have a large armamentarium of plants in their pharmacopoeias which are used throughout South East Asia. A similar situation exists in Africa and South America. Thus, a very high percentage of the world's population relies on medicinal and aromatic plants for their medicine. Western medicine is also responding. Already in Germany all medical practitioners have to pass an examination in phytotherapy before being allowed to practise. It is noticeable that throughout Europe and the USA, medical, pharmacy and health related schools are increasingly offering training in phytotherapy.

Multinational pharmaceutical companies have become less enamoured of the single compound magic bullet cure. The high costs of such ventures and the endless competition from me too compounds from rival companies often discourage the attempt. Independent phytomedicine companies have been very strong in Germany. However, by the end of 1995, eleven (almost all) had been acquired by the multinational pharmaceutical firms, acknowledging the lay public's growing demand for phytomedicines in the Western World.

The business of dietary supplements in the Western World has expanded from the Health Store to the pharmacy. Alternative medicine includes plant based products. Appropriate measures to ensure the quality, safety and efficacy of these either already exist or are being answered by greater legislative control by such bodies as the Food and Drug Administration of the USA and the recently created European Agency for the Evaluation of Medicinal Products, based in London.

In the USA, the Dietary Supplement and Health Education Act of 1994 recognised the class of phytotherapeutic agents derived from medicinal and aromatic plants. Furthermore, under public pressure, the US Congress set up an Office of Alternative Medicine and this office in 1994 assisted the filing of several Investigational New Drug (IND) applications, required for clinical trials of some Chinese herbal preparations. The significance of these applications was that each Chinese preparation involved several plants and yet was handled as a **single** IND. A demonstration of the contribution to efficacy, of **each** ingredient of **each** plant, was not required. This was a major step forward towards more sensible regulations in regard to phytomedicines.

My thanks are due to the staff of Harwood Academic Publishers who have made this series possible and especially to the volume editors and their chapter contributors for the authoritative information.

Roland Hardman

PREFACE

A distinctive smell in the dispensary of my father's pharmacy was my first acquaintance with *Valeriana*. I grew up in a small country town in the West of England in the Sixties during a period when the skills of pharmacists in extemporaneous preparation of medicines were being replaced by pre-packaged products and when many old plant-based remedies were giving way to more potent and effective drugs consisting of single chemical entities.

In the comparative backwater of the Cotswolds, some doctors still preferred the older remedies so my father often made up a large bottle of Potassium Bromide and Valerian Mixture which was prescribed as a tranquillizer or sedative. This entailed the use of Valerian Infusion, an alcoholic extract of *Valeriana officinalis* roots, with its distinctive and penetrating smell. Many find the odour repulsive but I, either through environmental or genetic conditioning, have never found it unpleasant.

At the same time, the benzodiazepines such as Valium ® and Librium ® were being introduced and used for much the same CNS depressant ends by those with a more modern inclination. Within ten years Valerian seemed to have disappeared, along with many other vegetable drugs, from mainstream pharmacy and its monograph was dropped from the 1973 British Pharmacopoeia.

This shift in medication did not occur to such a great extent in some other European countries such as Germany where herbal medication was still widely practised by general practitioners and so, in these countries, Valerian and other drugs were still used. Consequently, with the incorporation of European Pharmacopoeia monographs into the British Pharmacopoeia from 1980 onwards, Valerian once more featured in the entries.

The renaissance of the status of Valerian was also carried by the high annual growth, estimated at about 10% in most developed countries, in consumer preference for plant-based self-medication. This interest in phytotherapeutic agents, or 'herbal remedies', has occurred since about 1980 and has directed renewed clinical and scientific attention towards such products and their plant sources. There is increasing realisation that the traditional use of these materials often has a scientific basis but also that they may provide new leads for 'conventional' pharmaceuticals.

In time I learnt that the Valerian plant *Valeriana officinalis*, known to me as a British wild flower and as a substance in my father's pharmacy, was one menber of a genus of plants used throughout the world in most traditional medical systems for much the same purpose. The therapeutic efficacy of extracts of *V. officinalis*, together with other *Valeriana* species, in providing sedation, has been proven by many pharmacological and clinical experiments. Such extracts were extensively used during the two world wars for treatment of the condition known somewhat euphemistically as 'shell shock'. However, in common with many other phytotherapeutic agents, the chemical basis for this activity remained largely a mystery since no 'active ingredient' could be clearly identified. This hampered wider clinical usage since efficacy could not be guaranteed on the basis of chemical analysis.

The development since 1960 of more refined separation techniques and spectroscopic methods of structural elucidation, followed closely by tests for biological activity using small amounts and much more specific targets such as receptors, has resulted in a considerable amount of research on *Valeriana*. New types of chemical constituents have been discovered in this genus and the basis of their mode of action has to some extent been elucidated. It should be emphasised, however, that the story still awaits completion and, every year, several research papers appear in scientific journals dealing with different aspects of *Valeriana* species.

Valeriana is a good example of the problematic and beneficial aspects of the use of plant materials as pharmaceutical agents. Issues such as variation in chemical content and composition, the difficulties of producing crops of consistent composition with high amounts of the desired compounds and of analysing extracts make the production of a uniform product less easy to achieve than when a single chemical entity is concerned. The fact that no one compound, or even group of compounds, is responsible for the overall pharmacological effect underlies the problems faced in these contexts.

On the other hand, more interesting and positive features of phytotherapy are exemplified by *Valeriana* species. The overall effect is produced by a variety of types of chemical constituents with a range of relevant activities. A clinical condition such as sleeplessness may be due to one or more of several factors and a single chemical substance may not affect the underlying cause in a particular patient whereas one of the cocktail of compounds present in a preparation such as Valerian extract may be more likely to do so.

This book is intended to give an informative overview of the present knowledge of all aspects of the use, constituents and trade in *Valeriana* products. I trust that interest of its readers will be stimulated, not only in the fascinating properties of this particular medicinal genus, but also in the scientific study of medicinal plants as a whole.

Peter J. Houghton

CONTRIBUTORS

Jenö Bernáth
Department of Medicinal Plant
Production
University of Horticulture and Food
Industry
Villányi út 29/31
1114 Budapest
H-1502 Hungary

R. Bos
Department of Pharmaceutical Biology
University Centre for Pharmacy
Groningen Institute for Drug Studies
(GIDS), University of Groningen
Antonius Deusinglaan 2
9713 AW Groningen
The Netherlands

Anthony C. Dweck
Research Director
Peter Black Medicare
Peter Black Cosmetics & Toiletries
Southern Distribution Centre
White Horse Business Park
Aintree Avenue, Trowbridge
Wilts. BA14 0XB
UK

Richard Foss
Agros Associates
Yew Tree House
School House Lane
Aylsham
Norfolk NR11 6EX
UK

Josef Hölzl
Institut für Pharmakologie und
Toxikologie der Philipps Universität
3550 Marburg
Germany

Peter J. Houghton
Pharmacognosy Research Laboratories
Department of Pharmacy
King's College London
Manresa Road
London SW3 6LX
UK

J.J.C. Scheffer
Division of Pharmacognosy
Leiden/Amsterdam Center for Drug
Research (LACDR)
Leiden University
Gorlaeus Laboratories
PO Box 9502
2300 RA Leiden
The Netherlands

H.J. Woerdenbag
Department of Pharmaceutical Biology
University Centre for Pharmacy
Groningen Institute for Drug Studies
(GIDS), University of Groningen
Antonius Deusinglaan 2
9713 AW Groningen
The Netherlands

1. AN INTRODUCTION TO VALERIAN VALERIANA OFFICINALIS AND RELATED SPECIES

ANTHONY C. DWECK

Research Director, Peter Black Medicare, Peter Black Cosmetics & Toiletries
Southern Distribution Centre, White Horse Business Park,
Aintree Avenue, Trowbridge, Wilts. BA14 0XB

CONTENTS

INTRODUCTION

The Importance of Valerian And *Valeriana*

A distinctive smell often pervaded pharmacies in Great Britain and several other countries before the advent of the modern benzodiazepine tranquillisers and similar drugs. This odour was due to the extracts of a drug, called in English 'Valerian', which was incorporated in a mixture with potassium bromide prescribed for patients needing a relief from over-excitement of the central nervous system (CNS). Valerian commonly used in northern European medicine is derived from the underground organs of *Valeriana officinalis* L, a member of the Valerianaceae. This plant is steeped in history and related species are used in traditional medicine in many other parts of the world. The original use of *V. officinalis* as a perfume or perhaps even as a source of food was totally different to its modern use in orthodox and herbal medicine as a sedative and calming agent. The stresses of late twentieth century industrial society have resulted in an increase of use and interest in alternative sedatives to those used in orthodox medicine. Many of the products intended for self-treatment of mild stress contain *Valerian* crude drug or extracts. The additional aspect of its being a drug still prescribed by some orthodox practitioners as well as phytotherapists mean that a large amount of this group of crude drugs are grown, processed and used each year. The most important commercial species are *V.officinalis, V. wallichii* (syn. *V. jatamansii*), *V. fauriei* and *V. edulis* and it is these species which are dealt with in this book.

A few comprehensive reviews dealing with the history, constituents and activity of Valeriana have been published in recent years (Houghton, 1988, 1994; Hobbs, 1994; Jasperson-Squibb,1990). *Valeriana* species have been used for many years and the history of their use, or the reported ethnobotany, often seems implausible and the style and description of the writing is sometimes quaint and seems amusing. However, it is often startling to discover, that in reviewing the data retrospectively one finds an 'Old Wives Tale' that has been given scientific proof. Galen, who by luck or by judgement, intimated that Valerian was a sedative, might have been somewhat bemused to learn that his findings took another fourteen centuries to be rediscovered.

The Smell

The smell has been described as being that of "tom" cats (Hooper, 1984), as dirty socks (Keville, 1991), as warm and camphorous (Hutchens, 1992), as a strong penetrating, disagreeable odour with a camphorous, bitter taste (British Pharmaceutical Codex 1923). Others say it is nauseous and unpleasant (Baraicli Levy, 1991) and it is for this reason that one of its old names was 'Phu' (Hobbs, 1994), a name interpreted from the explanation of disgust with the strong smell of long-dried valerian root. He also likens the odour to well-seasoned dirty socks, while another author goes as far to describe it as

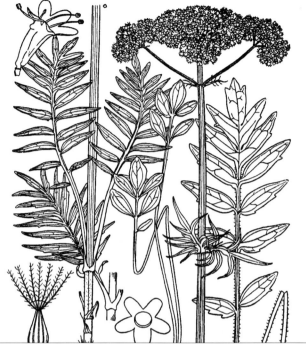

Figure 1

vile smelling (Saunders, 1976). Another (Heinerman, 1988) even describes the smell as 'unwashed underwear". Fresh valerian root smells like ancient leather but, when dried, it is nearer to stale perspiration (Bremness,1988).

However, this dislike of the smell was not the case in the years up to and including the 16th century, when most reports make no adverse comments on the odour of the plant. In the East the smell was liked immensely and used as a perfume in bathing. Today the oil of valerian is used as a component of many blended perfume oils, where it can impart a 'leathery' note to a fragrance. Valerian is not a recognised aromatherapy oil, and yet there is a growing body of evidence to show that the odour of valerian alone is sufficient to have a sedative effect (Balacs,1992).

The intense smell can be a problem and contaminate equipment, but the odour may be removed from a scale pan or from the hands by rubbing with sodium bicarbonate (Murrison,1935). It is now known that the major part of the odour is due to the isovaleric acid released by enzyme hydrolysis from some of the compounds present in the plant.

The Name

There are a number of explanations for the origin of the name but no universally-accepted etymology as yet exists. It might be from the Latin 'valere' meaning courage, which one might need to drink the infusion. Another source suggests from the Latin

Figure 2

'valeo' to be strong, or from 'valere' meaning to be in good health or to be healthy (Coombes,1985). The name valerian is also thought by some to be derived from the latinised form of the Old High German 'Baldrian' whose origin is uncertain.

BOTANICAL ASPECTS OF VALERIANA

Description of Genus

Valeriana is the major genus of the Valerianaceae, a family represented in all the temperate and sub-tropical areas of the world. Bentham and Hooker (1954) describe the Valerianaceae as being herbs with opposite leaves and no stipules. The flowers are in terminal corymbs or panicles and are usually small and numerous. The calyx is above the ovary and forms a small, sometimes toothed, border which is scarcely noticeable at the time of flowering but afterwards unrolls into a feathery pappus. The corolla is usually monopetalous, tubular at the base with five spreading lobes.. There are less stamens than lobes of the corolla. The fruit is small, dry and seed-like with a single seed suspended from the top of the cell but also frequently one or two imperfect or abortive empty cells. The genus *Valeriana* is characterised by perennials having three stamens and no spurs on the slightly swollen base of the corolla. The fruit is crowned by a feathery pappus.

Table 1 continued   Species of Valeriana and their common names

Species name	Synonyms	Common names
V. acutiloba Rydb.	*V. dioica*	Tobacco root, Wild heliotrope, Ill Cordileran Valerian, Downy fruit Valerian
V. adscendens Turcz.		
V. agrimonifolia Killip		
V. alliariifolia Troitzky		
V. alpestris Stev.		
V. alternifolia Ledeb	*V. officinalis* L	
V. amphilophis Graebn.		
V. amurensis P. Smirn. ex Kom..		
V. angustifolia Tausch	*V.officinalis*	Japanese Valerian, Kesso
V. arizonica A.Gray		Arizona Valerian
V. armeriifolia Schltdl.	*V. coarctata* Ruiz & Pav.	
V. aschersoniana Graebn. ex Weberbauer	*Aretiastrum aschersonianum* Graebn.	
V. asplenifolia Killip		
V. baltana Graebn.		
V. bambusicaulis Killip		
V. candolleana Gardner		
V. capitata Pall		Clustered Valerian
V. celtica L.	*Nardus celtica*	Celtic Nard
V. cephalantha Schlecht.		
V. chaerophylla Pers		
V. chaerophylloides Sm.		
V. clematitis Kunth		
V. clematoides Graebn.		
V. coarctata Ruiz & Pav		
V. coccinea = *Centranthus ruber*		
V. comosa Eriksen		
V. condamoana Graebn.		
V. connata Ruiz & Pav		
V. connata var. *nutans* Graebn		
V. convallarioides (Schmale) B.B. Larsen	*Phyllactis convallarioides* Schmale	
V. coreana Briq.		African Valerian
V. cornucopiae L.		
V. costata Schmale		
V. crassipes (Wedd.) Hoeck		
V. cumbemayensis Eriksen		
V. decussata Ruiz & Pav.		
V. dioica L.	*V. acutiloba*	Marsh Valerian, Tobacco root, Wild Heliotrope
V. Dioica var. *Sylatica*		Marsh Valerian
V. dioscorides	*V. phu*	Phu

Table 1 continued   Species of Valeriana and their common names

Species name	Synonyms	Common names
V. dipsacoides Graebn.		
V. edulis Nutt. ex. Torr. & Gray	*V. mexicana*	Tobacco Root, Edible Valerian
V. edulis Nutt, ssp *procera* Meyer		Mexican Valerian, Tobacco Root
V. elatior Graebn.		
V. erysimoides Poepp. & Endl.		
V. exaltata Mikan		
V. excelsa Poiret	*V. officinalis* (2n=56)	
V. exscapa Griseb.		
V. fauriei Briq.		English: Japanese Valerian, Kisso
		Japanese: Kanokosou, Kissoukon
V. friasana Schmale		
V. globifera Pers.	*V. globiflora* Ruiz & Pav.	
V. globularioides Graebn.		
V. globularis A. Gray	*V. globularis* Graebn	
V. grisiana Wedd.	*V. grisiana* Killip	
V. hadros Graebn.		
V. hardwickii Wall.		
V. herrerae Killip		
V. hirsutissima Killip	*V. macbridei* Killip	
V. hirtella Kunth		
V. hispida (Wedd.) Hoeck	*V. niphobia* Briq.	
V. hortensis Lam.		
V. humboldtii Hook. & Arn.	*V. humboldtii* Briq.	
V. hyalinorhiza Ruiz & Pav.	*V. mathewsii.*Briq.	
V. imbricata Killip	*Aretiastrum imbricatum* (Killip) Killip	
V. interrupta Ruiz & Pav.		
V. interrupta Ruiz & Pav. var. *interrupta*	*V. interrupta interrupta* (Ruiz & Pav.) Dufresne *V. interrupta interrupta* Graebn *V. interrupta* var. *interrupta* Ball	
V. interrupta var. *elatior* (Graebn.) Killip	*V. interrupta elatior* Graebn	
V. interrupta var. *minor* Ball	*V. interrupta* Ruiz & Pav. var. *interrupta*	
V. isoetifolia Killip		
V. italica		Phu, Gallic Nard
V. jasminoides Briq.		
V. jatamansii Jones	*V. wallichii* DC	Indian Valerian, Nard, Spikenard, Sunbul, Syrian Nard, Indian Nard
V. kilimandascharia Engl.		
V. laciniata Ruiz & Pav	*Astrephia chaerophylloides* (Sm.) DC.	
V. laevigata Willd. ex Steud.	*V. humboldtii* Hook. & Arn.	

Table 1 continued Species of Valeriana and their common names

Species name	Synonyms	Common names
V. ledoides Graebn.		
V. linearifolia Killip	*V. grisiana* Wedd.	
V. locusta L.		Common Corn Salad, Lamb's Lettuce
V. longifolia Kunth		
V. longifolia var. *pilosa* (Ruiz & Pav.) Wedd	*V. pilosa* Ruiz & Pav.	
V. lyrata M. Vahl	*V. lyrata* Ball *V. lyrata* Graebn	
V. macbridei Killip		
V. malvacea Graebn.		
V. mathewsii Briq.	*V. humboldtii* Hook. & Arn.	
V. maxima Killip	*V. dipsacoides* Graebn.	
V. melanocarpa Killip	*V. decussata* Ruiz & Pav.	
V. merxmuelleri Seitz		
V. mexicana DC	*Centranthus ruber* *V. sorbifolia* H.B.K. var. *mexicana* (DC) F.G. Mey.	
V. microphylla Kunth	*V. microphylla* Graebn	
V. micropterina Wedd		
V. montana Bieb	*V. alpestris*	
V. nigricans Graebn.		
.*V. niphobia* Briq.	*V. niphobia* (Wedd.) Graebn	
V. nitida Kreyer		
V. nivalis Wedd.		
V. oblongifolia Ruiz & Pav.		
V. obtusifolia DC.	*V.obovata*	
V. occidentalis		Small-flower Valerian
V. officinalis L..		Valerian, All heal, Amantilla, Belgian valerian, Capon's Tail, Cat's love, Cat's valerian, Common valerian, Fragrant valerian, Garden heliotrope (wrongly), Garden Valerian, Great wild valerian, Herba Benedicta, Ka-no-ko-so, Kesso root, Kissokon, Kuanyexiccao, Luj, Nard, Ntiv, Officinal Valerian Racine de Valeriane St. George herb, Setwall, Setwell, Theriacaria, txham laaj, Valerian fragrant, Great wild valerian, V. extranjera, V., Valeriane, and Vandal root, Waliryana, Wild valerian

Table 1 continued Species of Valeriana and their common names

Species name	Synonyms	Common names
		French: Valériane
		German: Baldrian, Katzenwurzel,
		Balderbrackenwurzel,
		Baldrianwurzel
		Italian: Valeriana
		Spanish: Valeriane
		Dutch: Valeriaan
		Welsh: Llysiau Cadwgan
V. officinalis L., subsp.*collina* Wallr.Nyman		
V. officinalis L. subsp. *sambucifolia* (Mik.f.) Celak		
V. officinalis var. *latifolia*	*V. angustifolia*	
V. officinalis var. *Mikanii*, Syme		English Valerian, Common valerian, Belgian Valerian, All Heal, Fragrant Valerian, Garden Valerian
V. oligodonta Killip	*V. plectritoides* var. *plectritoides* Graebn.	
V. oxyrioides Graebn.	*V. lyrata* M. Vahl	
V. paniculata Ruiz & Pav.		
V. pardoana Graebn.		
V. parvula Killip		
V. pauciflora Michx.		Large-flower Valerian
V. pavonii Poepp. & Endl.		
V. pedicularioides Graebn.	*V. interrupta* Ruiz & Pav. var. *interrupta*	
V. pennellii Killip		
V. phu L.		
V. pilosa Ruiz & Pav.	*V. pilosa* (Ruiz & Pav.) Wedd.	
V. pimpinelloides Graebn.	*V. interrupta* var. *elatior* (Graebn.) Killip	
V. pinnatifida Ruiz & Pav		
V. plectritoides Graebn.		
V. plectritoides var.*pallida* Graebn.		
V. plectritoides var. *plectritoides* Graebn.	*V. plectritoides plectritoides* Killip	
V. pratensis	*Phyllactis pratensis*	
V. procurrens Wallr.		
V. pulchella Mart. et Gal.		
V. punctata F. Meyer		
V. pycnantha A. Gray		
V. pygmaea Graebn.	*V. globularis* A. Gray	
V. pyrenaica L.		Pyrenean Valerian

Table 1 continued   Species of Valeriana and their common names

Species name	Synonyms	Common names
V. quadrangularis Kunth		
V. radicata Graebn.		
V. remota Ball	*V. lyrata* M. Vahl	
V. renifolia Killip		
V. repens Host.	*V. procurrens* Wallr.	
V. rhizantha A. Gray	*Stangea rhizantha* (A. Gray) Killip	
V. rigida Ruiz & Pav.	*Phyllactis rigida* (Ruiz & Pav.) Pers.	
V. romanana Graebn.	*V. condamoana* Graebn.	
V. rufescens Killip	*V. rumicoides* Wedd.	
V. salina Pleigel		
V. sambucifolia Mikan.		
V. saxatilis L.		
V. scandens L.		Florida Valerian
V. scouleri		Scouler's Valerian
V. serrata Ruiz & Pav.		
V. sitchensis Bong.		Sitra Valerian
V. sitchensis sp. *scouleri* Bong.		Pacific Valerian
V. sorbifolia H.B.K. var. *mexicana* (DC) F.G. Mey.		
V. spathulata Ruiz & Pav.	*Belonanthus spathulatus* (Ruiz Lopez & Pavon) Schmale	
V. sphaerocephala Graebn.		
V. stolonifera Czern.		
V. stubendorfi Kreyer ex Kom		
V. supina L.		
V. sylvatica Banks	*V. dioica*	
V. sylvestris Grosch.	*V. officinalis* L.	
V. tenuifolia Ruiz & Pav.	*Phyllactis tenuifolia* (Ruiz & Pav.) Pers.	
V. tessendorffiana Graebn.		
V. texana		Guadalupe Valerian
V. thalictroides Graebn.		
V. tiliifolia Troitzky		
V. tomentosa Kunth		
V. trichomanes Graebn.		
V. tripteris L.		Three-leaved valerian
V. tuberosa L.	*Nardus montana*	Valeriane
V. urticifolia Kunth		
V. variabilis Graebn.		
V. verrucosa Schmale		
V. versifolia Brügger		
V. virgata Ruiz & Pav.		
V. wallichii DC	*V. villosa* Wall.	English: Indian Valerian.

Table 1 continued   Species of Valeriana and their common names

Species name	Synonyms	Common names
	V. jatamansi Jones *V. spica* Vahl	Ayurveda: Tagar Sinhalese: Kattakumanjal, Thuwarda Sanskrit: Barhena, Chakra, Danda, Dandahasta, Dipana, Hasti, Jimha, Kalanusakara, Kalanusariva, Kalanusarya, Kshatra, Kunchina, Kutil, Laghusha, Mahoraga, Nahushakhya, Nata, Padika, Parthiva, Pindatagara, Rajaharshana, Shatha, Tagara, Vakra, Vinamra.
V. warburgii Graebn. *V. weberbaueri* Graebn.		

Table 2   Important plants related to Valeriana

Plant species	Common name
Centranthus macrosiphon Boiss	Long-spurred Valerian
Centranthus ruber (L.) DC.	English: Red Valerian, Jupiter's Beard, Fox's Brush Welsh: Triaglog Coch
Fedia cornucopiae Gaertn.	English: Horn-of-Plenty, Valerian
Nardostachys jatamansi DC.= *Nardostachys grandiflora* DC.= *Patrinia* *jatamansi* Don, = *Valeriana jatamansi* Wall. = *Fedia grandiflora* Wall.	English: Spikenard Sinhalese: Jatamansa Tamil: Jatamashi, Kanuchara Hindi: Balchhar, Balchir, Baluchar, Jatalasi, Jatamansi, Kanuchara Sanskrit: Akashamansi, Amritajata, Bhutajata, Bhutakeshi, Chakravartini, Gandhamansi, Gauri, Hinsra, Jadamansi, Janani, Jatala, Jatamansi, Jatavali, Jati, Jatila, Keshi, Khasambhava, Kiratini, Kravjadi, Krishnajata, Laghumansi, Limasha, Mansi, Mansini, Mata, Mishika, Misi, Mrigabhaksha, Nalada, Nirlamba, Parvatavasini, Peshi, Peshini, Pishachi, Pishita, Putena, Sevali, Shvetakeshi, Sukshmajatamansi, Sukshmapatri, Tamasi, Tapasvini, Vahnini.
Patrinia scabiosaefolia Fisch	
Valerianella carinata Loisel	Keeled-fruited Corn Salad
Valerianella coronata DC.	
Valerianella dentata (L.) Pollich. = *V.* *morisonii*	Narrow-fruited Corn Salad

Table 2 Continued

Plant species	Common name
Valerianella eriocarpa Desv	Hairy-fruited Corn Salad, Italian Corn Salad
Valerianella locusta L. Latterade	Common Corn Salad, Lamb's Lettuce
Valerianella olitoria Pollich	Lamb's Lettuce or Corn Salad
Valerianella radiata	Bearded Corn Salad
Valerianella stenocarpa	Narrow-cell Corn Salad
Valerianella umbilicata	Navel Corn Salad
Valerianella rimosa Bast. = *V. auricula*	Broad-fruited Corn Salad

Table 3 Plants not related to valeriana spp. but with similar names

Plant species	Common name
Cypripedium pubescens Willd. (Orchidaceae)	American Valerian, Ladies' Slipper
Cypripedium acaule Ait. (Orchidaceae)	American Valerian, Pink Lady's Slipper
Cypripedium calceolus var. *pubescens* Correll(Orchidaceae)	American Valerian
Heliotropium arborescens (Boraginaceae)	Garden Heliotrope
Polemonium coeruleum L. (Polemiaceae)	Greek Valerian, Jacob's Ladder
Polemonium reptans L. (Polemiaceae)	Greek Valerian, Jacob's Ladder
Senecio aureus (Asteraceae)	Life Roots, False Valerian

Valeriana Species

The number of species of valerian is fairly large and is listed in Table 1 together with synonyms and common names where applicable.

Related Plants of Valerianaceae Of Importance

Although this book is concerned with the genus *Valeriana* it is important to realise that other members of the Valerianaceae are also important, particularly as sources of medicinal, cosmetic and food materials. A list of the most important of these is given in Table 2.

Non-Related Plants With Similar Trivial Names

As with all plants that share common names, there is always the possibility that people will become confused and use the incorrect plant, trusting in the vernacular name alone. The list shown in Table 3 is a compilation of the most commonly reported errors and mistakes.

HISTORICAL AND MODERN USES OF VALERIAN

This aspect of the information concerning Valerian begins with all the strange and mystical beliefs that surround it and a collection of anecdotes and folk lore that might otherwise be difficult to find assembled in one place.

Valerian]a In Folklore

Miscellaneous beliefs

The following section is a short list of miscellaneous statements that have been collected over the years. They are given without comment, because science has a wonderful knack of occasionally proving or giving respectability to even the most unlikely of events.

A young woman who carries a sprig of valerian is never said to lack ardent lovers.

It was said to inspire love and so was used as an ingredient in love philtres. It was also cited as an aphrodisiac (Gordon,1980). It has also been said that valerian increases psychic perception (Howard,1987).

It protects a person from thunder and lightening, and was used both for and against witchcraft (Gordon,1980).

Hanging the herb in the house was reputed to prevent the husband and wife from bickering (Law, 1973).

Excessive dependance on valerian is said to cause headaches, mental agitation, much restlessness and severe cases of delusion. It is said that Adolf Hitler was a valerian addict and regularly took large and excessive doses (Bairacli Levy, 1991).

Planetary influences

There was a time when almost everything was considered to come under the influence of the planets and astral bodies. The astrologers and mystics assigned the futures of the people and the future of the world according to the positioning of those stars and heavenly configurations. Trees, precious stones and even plants came under the influence of the firmament. According to Law (1973) valerian is under the influence of Uranus, however, Culpeper (Potterton, D. (ed) 1983) and Gordon (1980) say that the plant is under the influence of Mercury.

Language of flowers

In medieval times, the art of writing was limited to only a few, and even then it was more common to write in Latin than in the common tongue. In the Middle Ages there developed a system of communication that revolved around plants (a custom which was revived in Victorian times and at the same time the list probably enlarged). In all likelihood, this floral symbolism started with the heraldic symbols used to distinguish knights in battle, and then developed into a floral language, for example:-

Borage:	'Your attentions only embarrass me'
Chamomile:	'I admire your courage, do not despair'
Pink Clover:	'Do not trifle with my affections'

Individual plants could be combined with a wide diversity of others in order to build up a quite a comprehensive message, and thus it became customary to present bouquets of flowers instead of writing letters or notes, and for this purpose hundreds of flowers had designated meanings.

There can be few who do not know the symbolic meaning of the red rose. However, be tempered with caution, for a yellow rose means 'misplaced affection' or "I love another"! In the book Language of Flowers (Anon: 1968), valerian is said to mean an accommodating disposition, which is confirmed in another text (Conway, 1975) which says that it means a concealed merit - "though lowly, I aspire to love you".

Saintly assignation

At more or less the same time that the astrologers were assigning all the plants to planetary influences, there were others who were dedicating the plants to various saints. Valerian is dedicated to St. Bernard (Gordon, 1980).

Valeriana In History

In a concise history of valerian, Hobbs (1994) pointed out that the early uses of valerian were in the most part for its bitter and aromatic qualities. Related plants, especially spikenard derived from *Nardostachys jatamansii*, are mentioned in the Bible.

> While the king sitteth at his table, my spikenard sendeth forth the smell thereof. **Song of Solomon I v.12**

> Thy plants are an orchard of pomegranates, with pleasant fruits; camphire, with spikenard and saffron; calamus and cinnamon, with all trees of frankincense; myrrh and aloes, with all the chief spices: **Song of Solomon IV v.13-14**

> Then took Mary a pound of ointment of spikenard, very costly, and anointed the feet of Jesus, and wiped his feet with her hair: and the house was filled with the odour of the ointment. **John XII v.3**

> And being in Bethany in the house of Simon the leper, as he sat at meat, there came a woman having an alabaster box of ointment of spikenard very precious; and she brake the box, and poured it on his head. **Mark XIV v.3**

The Greek physician and pharmacist, Galen (131-201 A.D.), was probably the first to allude to the sedative qualities of valerian. However, it was not until the end of the 16th. century that this use was truly recognised. The virtues of Setwall (the common name at that time for Valerian) were described by Gerard (1597) as follows:-

> The dry root is put into counterpoysons and medicines preservative against the pestilence, as are treacles, mithridates, and such like: whereupon it hath been had (and is to this day among the poore people of our Northerne parts) in such veneration amongst them, that no broths, pottage or physicall meats are worth any thing, if Setwall were not at an end: whereupon some women Poet or other hath made these verses.

> > They that will have their heale,
> > Must put Setwall in their keale.

> It is used generally in sleight cuts, wounds, and small hurts

Slightly later, another herbalist (Culpeper, 1653) records:

> This is under the influence of Mercury. Dioscorides saith, That the Garden Valerian hath a warming faculty, and that being dried and given to drink it provokes urine, and helps the stranguary. The decoction thereof taken, doth the like also, and takes away pains of the sides, provokes women's courses, and is used in antidotes. Pliny saith, That the powder of the root given in drink, or the decoction thereof taken, helps all stoppings and stranglings in any part of the body, whether they proceed of pains in the chest or sides, and takes them away. The root of Valerian boiled with liquorice, raisins, and anniseed, is singularly good for those that are short-winded, and for those that are troubled with the cough, and helps to open the passages, and to expectorate phlegm easily. It is given to those that are bitten or stung by any venemous creature, being boiled in wine. It is of a special virtue against the plague, the decoction thereof being drank, and the roots being used to smell to. It helps to expel the wind in the belly. The green herb takes away the pains and prickings there, stays rheum and thin distillation, and being boiled in white wine, and a drop thereof put into the eyes, takes away the dimness of the sight, or any pin or web therein. It is of excellent property to heal any inward sores or wounds, and also for outward hurts or wounds, and drawing away splinters or thorns out of the flesh.

Valerian was first used therapeutically as a sedative by the English doctor John Hill in the middle of the 18th century. About a hundred years later this medical plant was described by Christoph Wilhelm Hufeland (1762 - 1836), the founder of the electric medical school and doctor to many important people, as the "best medicine for the nerves". The medical drug valerian was chemically analysed already in the 19th century, the volatile oil in particular being the object of various investigations. By the end of the century the pharmaceutical texts (Dispensatory of the United States of America: 1883) were starting to describe the uses of valerian in a way that would almost be accepted today.

> Valerian is gently stimulant, with an especial direction to the nervous system, but without narcotic effects. In large doses it produces a sense of heaviness and dull pain in the head, with various other effects indicating nervous disturbance. The oil, largely taken, is said by M. Barailer, from his own observation, to produce dullness of intellect, drowsiness ending in deep sleep, reduced frequency of pulse, and increased flow of urine. It is useful in cases or irregular nervous action, when not connected with inflammation, or an excited condition of the system. Among the complaint in which it has been particularly recommended are hysteria, hypochondriasis, epilepsy, hemicrania, and low forms of fever, attended with restlessness, morbid vigilance, or other nervous disorder. It has also been used in intermittents, combined with Peruvian bark, and in acute rheumatism. As the virtues of valerian reside chiefly in the volatile oil, the medicine should not be given in decoction or extract. The distilled water is used on the continent of Europe; and the volatile oil is occasionally substituted with advantage for the root.

The roots of *Valeriana dioica* are said to be sometimes mingled with those of the officinal plant; but the adulteration is attended with no serious consequences; as, though much weaker than the genuine valerian, they possess similar properties.

> By 1923, there was the first indication (British Pharmaceutical Codex: 1923) that the action of valerian could also act through an odorous pathway. The action of valerian rhizome is virtually that of its volatile oil, the valerianic esters of which have no stimulating action on the physical functions and the circulation, as was formerly believed, although they possess the usual carminative action of the volatile oils. The action of such malodorous substances as valerian is generally attributed to the mental effect produced by their unpleasant odour and taste. Valerian is used as a carminative and antispasmodic in hysteria and similar nervous manifestations.

Valerian has been prescribed (Howard, 1987) as the perfect herbal tranquilliser, and was used for this purpose in the First World War to treat soldiers suffering from shell shock (Howard, 1987).

During the Second World War there was a shortage of the dried rhizome and roots of *Valeriana officinalis*, collected in the autumn and a special dispensation was printed in the 1941 edition of Martindale: "*As a war emergency measure, when valerian is prescribed or demanded, Indian valerian may be dispensed or supplied.*" The description of the uses read as follows:

> Given in hysterical and neurotic conditions as a sedative. Its action has been attributed to its unpleasant smell, and if this is so, deodorised preparations cannot possess any activity due to their valerian content.

Modern Medicinal Uses Of Valerian

Valerian in Western orthodox medicine and phytotherapy

Valerian today is a highly respected medicinal plant with many Pharmacopoeial monographs. (Newall et al., 1996). Entries can be found in the current British Pharmacopeia, European Pharmacopeia, British Herbal Pharmacopoeia 1990, British Pharmaceutical Codex 1963, Martindale 30th edition and the pharmacopeias of Austria, Brazil, Czechoslovakia, Egypt, France, Germany, Greece, Hungary, Italy, Netherlands, Norway, Romania, Russia, Switzerland and Yugoslavia. The Egyptian Pharmacopoeia mentions valerian from *Valeriana wallichii* (Indian Valerian). Japan has Japanese Valerian from *Valeriana fauriei* or allied plants

The British Herbal Pharmacopoeia: (1983) describes Valerian as a sedative, mild anodyne, hypnotic, spasmolytic, carminative and hypotensive indicated for hysterical states, excitability, insomnia, hypochondriasis, migraine, cramp and rheumatic pain. Another source (Weiss, 1986) says that the three main areas of use for valerian are for nervous excitement, nervous sleeplessness, and nervous palpitations.

A considerable amount of work has been carried out to determine the compounds responsible for this activity and to elucidate their pharmacological effects. Extracts and their constituents have been shown to exert smooth muscle relaxant, sedative, antidepressant and sleep-inducing properties. These are discussed in more detail in chapter 3.

Medicinal Uses Of Valeriana In Other Cultures

In India, *Nardostachys jatamansi* DC. Mem. is used (Jayaweera, 1982). The roots of this are supposed to possess stimulant and antispasmodic properties. They are used in the treatment of epilepsy, hysteria, convulsive ailments, palpitations of the heart, consumption, diseases of the eye, itch, boils, swellings, diseases of the head, hiccough, etc. Also mentioned is *Valeriana wallichii* DC. Mem., where the root is used as one of the ingredients in the preparation of snake bite cures. It is also used for liver, kidney and spleen diseases.

In Chinese medicine (Leung and Foster: 1996), both common and Indian valerian as well as those of *Valeriana coreana* Briq., *V. stubendorfi* Kreyer ex Kom., *V. amurensis* P. Smirn. ex Kom., and *V. hardwickii* Wall. are similarly used. In addition, they are use in treating chronic backache, numbness due to rheumatic conditions, colds, menstrual difficulties, bruises and sores etc., among others, generally as a decoction or alcoholic infusion. *Valeriana sylvatica* was found in the medicine bag of Canadian Indian warriors as a wound antiseptic (Bremness, 1988).

Cosmetic Uses

Bathing

Valerian has been used for perfume baths in the East (Graves, 1990) and also as a soothing herb bath (Weiss, 1986) or as a vapour bath (Hutchens, 1992). According to another source (Bremness, 1991) a decoction is used as a facial wash.

Topical application

The use of an alcoholic extract of valerian for the treatment of dandruff has been mentioned (Leung, 1980) and also the use of valerian for treating sores and pimples. It is said (Bremness, 1988) that the use of a lotion is good for the treatment of acne and skin rashes.

A report of the Council of Europe (1989) on the intended cosmetic effect recommended a maximum concentration of up to 2% of the essential oil in cosmetic products. The root and rhizome are used. They report that valerian is soothing, hampers sweat secretion and is relaxing. It is a fragrance material due to its essential oil. They list other possible effects as sedative, hypotensive, antispasmodic, analgesic and anti-inflammatory. In another report (Jayaweera, 1982) it is said that valerian is used for ailments of the hair. The roots are also used for improving the complexion, increasing the lustre of the eye and promoting the growth and increasing the blackness of the hair.

Food Uses

Valeriana and related species are used as minor food plants in some areas. They might not be to the taste of everyone, however, throughout history, there are reports of the plant being used as a source of food. *Valeriana cornucopiae* Linn. or African Valerian, is a native of the Mediterranean region and provides a salad plant. *Valerian edulis* Nutt. or Tobacco Root, is the principle edible root amongst the Indians who inhabit the upper waters and streams on the western side of the Rocky Mountains (Hedrick, 1972). In another text Saunders (1976) refers to it as another staple of some tribes, occurring in damp grounds from the Great Lakes to Oregon and British Colombia. Its deep, perpendicular root is vile smelling and ill tasting, but long steaming makes it palatable, at least to the Indians. Fremont speaks well of it in his journal, under the Snake name kooyah, though his associate Preuss could not stay in the same tent with it, much less eat it.

Centranthus macrosiphon Boiss. (Long-spurred Valerian) is used as a salad plant, particularly in France and the related *Centranthus ruber* (L.) DC. is eaten as a salad (Mabey, 1972) in southern Italy and France. The young leaves of this plant are sometimes boiled with butter as greens, or eaten raw in salads, though they can be rather bitter used in this way. *Fedia cornucopiae* Gaertn. or Horn-of-Plenty Valerian, is grown in France as a salad plant as is *Valerianella coronata* DC.

Valerianella eriocarpa Desv. or Italian Corn Salad, is much used in Europe as a substitute for lettuce in the spring, and when grown in rich soil as a substitute for spinach. *Valerianella locusta* or Lamb's Lettuce, Common Corn Salad. is an original European species, native to the Mediterranean (Lanska, 1992). It grows wild throughout Europe, in the Near East, Caucasus, Northern Africa and North America. According to some it is a spring salad vegetable with a pleasant, slightly nutty flavour. It contains about 60 mg of vitamin C, and a large amount of phosphorous, calcium, iron, saccharides, proteins, fats etc. Lamb's Lettuce is used only fresh in mixed and potato salads, in herb soups and omelettes. *Valerianella olitoria* Pollich, commonly Lamb's Lettuce or Corn Salad, one of the widest European examples of a valerian salad herb.

Effect Of *Valeriana* On Animals

There are frequent references to animals and their reaction to valerian. It is said that cats are known to be attracted by valerian, which can cause a state of ecstasy in them (Weiss, 1986). This is reported by a number of authors (Phelps-Brown, 1993; Spoerke, 1990; Blackmore, 1982) to name but a few. It is also said (Graves, 1990) that the male is more affected by the plant than the female and that cats are so fond of the herb that they will dig up the roots and eat them (Hutchens, 1973).

The root also excites rats and that the famous Pied Piper of Hamlyn may not have been such a great flute player, but more of a success because of the valerian root that he carried in his pockets! (Keville, 1991), a thought echoed by other authors (Howard, 1987; Ceres: 1984).

Gypsies use the herb to quieten unfriendly dogs, and horses are also attracted by the smell of valerian (Conway,1975).

Finally, it is said (Bremness, 1988) that the whole plant stimulates growth of nearby vegetables by stimulating phosphorus and earthworm activity, since the roots also attract the worms.

REFERENCES

Anon. (1992) *The Language of Flowers*. 13th impression Penguin Books Ltd, London.

Bairacli Levy, J.de. (1991) *The illustrated herbal handbook for everyone*. 4th edition. Faber and Faber, London.

Balacs, T. (1992) Research Reports. *The International Journal of Aromatherapy*, **4**, 28–30.

Bentham , G. and Hooker, J.D. (1954) *Handbook of the British Flora* 7th ed. (rev. A. B. Rendle) Reeve, Ashford, UK, p. 224.

Blackmore, S. (1982) *The Illustrated Guide to Wild Flowers*. Blitz Editions, London.

Bremness, L. (1988) *The Complete Book Of Herbs*. CLB (Colour Library Books) Dorling Kindersley, London.

British Herbal Pharmacopoeia. (1983) B.H.M.A., Bournemouth, UK.

British Pharmaceutical Codex (1923) The Pharmaceutical Press, London.

Ceres (1984) *The Healing Power of Herbal Teas*, Thorsons, London.

Conway, D. (1975) *The Magic of Herbs*. Jonathon Cape, London.

Coombes, A.J. (1985) *Dictionary Of Plant Names*. Hamlyn, London.

Council of Europe (1989) *Plant Preparations Used As Ingredients Of Cosmetic Products. 1st Edition*. Strasbourg. HMSO, London.

Culpeper, N. (1653) *Culpeper's Complete Herbal - a book of natural remedies for ancient ills*. Wordsworth Reference Editions Ltd. (1995) London.

Dispensatory of the United States of America, (1883) 15th edition. J.B. Lippincott & Co., Philadelphia.

Gerard, J. (1597) *Gerard's Herbal*. ed. Marcus Woodward. Studio editions. (1990), London.

Gordon, L. (1980) *A Country Herbal*. Webb and Bower Ltd., London.

Graves, G. *Medicinal Plants* - 1990 Bracken Books, London.

Hedrick, U.P. (editor). (1972) *Sturtevant's Edible Plants of the World*. Dover editions, New York.

Heinerman, J. (1988) *Heinerman's Encyclopedia of Fruits, Vegetables and Herbs*. Parker Publishing Company, New York.

Hobbs, C. (1994). *Valerian - the relaxing and sleep herb*. Botanica Press, Capitola, California, USA.

Holy Bible, The King James I. version

Hooper, M. (1989) *Herbs and Medicinal Plants.* Kingfisher Books, London.

Houghton, P.J. (1988) The biological activity of Valerian and related plants. *J. Ethnopharmacol.* **22**, 121–142.

Houghton, P.J. (1994) Valerian. *Pharm. J.*, **253**. 95–96.

Howard, M. (1987) *Traditional Folk Remedies, A comprehensive Herbal.* Century , London.

Hutchens, A.R. (1992) *A Handbook of Native American Herbs.* Shambhala, Chicago.

Hutchens, A.R. (1973) *Indian Herbalogy of North America.* Shambhala, Chicago.

Jaspersen-Squib, R (1978). Sédatifs à base de plantes. *Schweiz. Apoth. Ztg.*, **118**, 503–508.

Jayaweera, D.M.A. (1982). *Medicinal Plants used in Ceylon,* Part 5. National Science Council of Sri Lanka, Colombo.

Keville, K. (1991) *The Illustrated Herb Encyclopaedia.* Grange Books, London.

Lanska, D. (1992) *The Illustrated Guide to Edible Plants.* Chancellor Press, London.

Law, D. (1973) *The Concise Herbal Encyclopaedia.* John Bartholomew and Son Ltd, London.

Leung, A.Y. (1980) *Encyclopedia Of Common Natural Ingredients Used In Food,Drugs And Cosmetics.* John Wiley, Chichester.

Leung, A.Y. and Foster, S. (1996) Encyclopedia of Common Natural Ingredients used in food, drugs and cosmetics. 2nd edition. John Wiley, Chichester.

Mabey, R. (1972) *Food for Free.* Collins, London.

Martindale. The Extra Pharmacopoeia. 29th. Edition. 1989. The Pharmaceutical Press, London.

Martindale. The Extra Pharmacopoeia. 22nd edition. 1941. The Pharmaceutical Press, London.

Murrison, R.G. *Pharm .J.,* (1935) 106.

Newall, C.A., Anderson, L.A. and Phillipson, J.D. (1996) *Herbal Medicines - a guide for health-care professionals.* The Pharmaceutical Press. London.

Phelps Brown, O. (1993) *The Complete Herbalist.* Newcastle Publishing (Van Nuys, California).

Potterton, D. (ed). (1983) *Culpeper's Colour Herbal.* W. Foulsham, London.

Saunders, C. F. (1976) *Edible and Useful Wild Plants of the United States and Canada.* Dover Books, New York.

Spoerke, D.G. (1990) *Herbal Medications.* Woodbridge Press, Santa Barbara, California.

Weiss, R.F. (1986) *Herbal Medicine.* The Bath Press, Bath, UK.

2. THE CHEMISTRY OF *VALERIANA*

PETER J HOUGHTON

Pharmacognosy Research Laboratories, Department of Pharmacy, King's College London, Manresa Road, London SW3 6LX

CONTENTS

INTRODUCTION

It is important to know the structure and reactions of the chemical compounds present in a crude drug, particularly those which contribute to its biological activity. Such knowledge arises, not only because of the innate curiosity of human beings, but also because it facilitates authentication of unknown material and also provides markers for quantitative evaluation.

A further benefit which may follow such knowledge is the introduction of a new chemical entity which may form the basis for designing semisynthetic drugs with the same activity but improved characteristics compared with the original molecule. The first two of these principles has been applied to some extent to Valerian but as yet only a little progress has been made regarding the third approach (Thies, Seitz and Moddelmog, 1984).

It is also important to know what compounds, other than those shown to be active, might be present in an active fraction since they may exercise a preservative role, improve bioavailability or solubility or act as a synergist of the major active substances present.

The medicinal importance of *Valeriana* species prompted initial chemical investigation quite early in the nineteenth century but little progress was made in determining the structure of the compounds present until the improvement in separation techniques and the introduction of spectroscopic methods of structural elucidation in the last half of the twentieth century. In hindsight it is not difficult to see why individual compounds were not isolated earlier since extracts and oils obtained from this genus have now been shown to consist of a complex mixture of closely-related substances, many of which are somewhat labile.

The traditional Northwestern European drug Valerian, the dried underground parts of *V. officinalis*, has a distinctive smell, which many people find abhorrent, and this is due to isovaleric acid **1** which is formed by enzymatic hydrolysis of some of the constituents during storage. Isovaleric acid was first isolated from *Valeriana* in 1963 (Schmeltz *et al.*) and multiples of its branched 5 carbon skeleton, called an isoprenoid, can be discerned in most of the important chemical compounds which have been isolated from *Valeriana* and related plants. In these compounds the carbon skeleton consists of 2 or 3 of these units, joined in a variety of ways, to form monoterpenoids and sesquiterpenoids respectively.

Such compounds are found as the major constituents of the volatile oils and also as less volatile constituents found in the cytoplasm of plant tissues.

The alkaloids found in this plant are also biogenetically related to the terpenoids and similarities in the carbon skeleton are evident.

This chapter will deal with the skeletal types and individual constituents found in Valeriana species with some emphasis on patterns of qualitative and quantitative variation and also on structure-activity relationships.

TERPENOIDS

The terpenoids arise as products of the mevalonate pathway and consist of multiples of branched 5-carbon units. The pathway is sometimes called the isoprenoid pathway since the branched 5-C unit has the isoprene skeleton. A 10-C unit is called a 'terpene' and the different types of structure are consequently designated monoterpenes, sesquiterpenes, diterpenes and triterpenes if they have a skeleton of 10, 15, 20 or 30 C atoms respectively.

An outline of the mevalonate pathway for the biogenesis of mono- and sesquiterpenoids is shown in Figure 1.

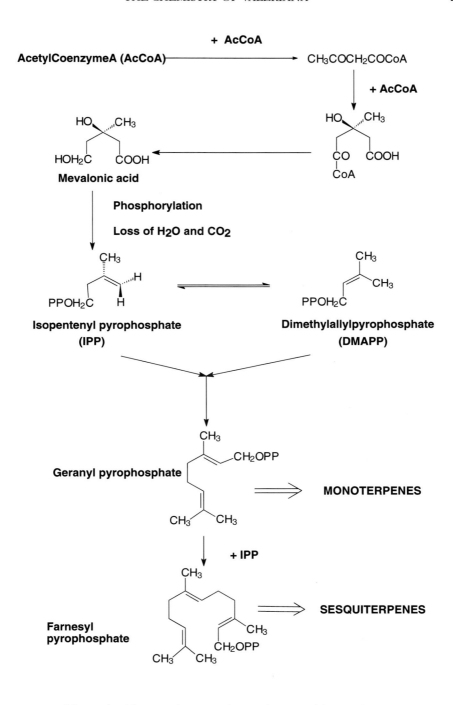

Figure 1   The mevalonate pathway of terpene biogenesis

Table 1 Side chain substituents found in Valeriana species

Structure	Abbr.	Name
CH_3CO-	Ac	Acetyl
(isovaleryl structure)	Iv	Isovaleryl
(α-acetoxyisovaleryl structure)	Aav	α-Acetoxyisovaleryl
(β-acetoxyisovaleryl structure)	Bav	β-Acetoxyisovaleryl
(3-methylcrotonyl structure)	Cr	3-Methylcrotonyl
(γ-acetoxycaproyl structure)	Aic	γ-Acetoxycaproyl
(β-methylisovaleryl structure)	Miv	β-Methylisovaleryl
(α-isovaleroyloxyisovaleryl structure)	Iiv	α-Isovaleroyloxyisovaleryl
(β-hydroxyisovaleryl structure)	Hiv	β-Hydroxyisovaleryl

Monoterpenes and sesquiterpenes with comparatively little oxygenation and substituent side-chains are often found in plants as a complex mixture known as a volatile oil. As the name implies, these oils have an appreciable vapour pressure, especially in highly humid conditions and in general are lipophilic rather than hydrophilic. They are often stored in specialised organs and many play an important role in the survival strategy of plant species.

A greater degree of oxygenation, nitrogenation, or combination with sugars to form glycosides is found in many terpenoids, especially the diterpenoids and triterpenoids. These compounds may be found in specialised cells or organelles but are also encountered as components of ordinary cytoplasm. Such types of terpenoids are not been reported present in *Valeriana*. However, in this genus, the terpenoids are often esterified with one or more acids (see Table 1)

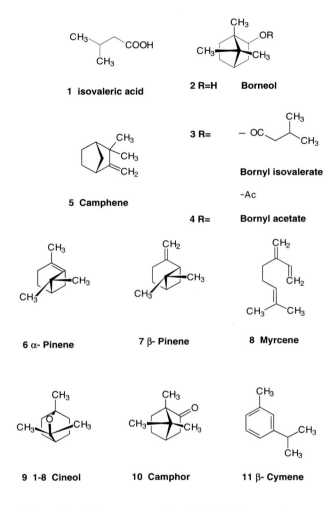

Figure 2 Monoterpenes found in Valeriana species

Volatile oils of *Valeriana* and their constituents

Most of the analysis of the volatile oil of Valeriana species has been performed on steam distilled oil but a recent study descibes the headspace analysis by linked gas chromatography-mass spectrometry of a drug sample (Nikiforov *et al.*, 1994).

The volatile oils produced from *Valeriana* consist of a mixture of mono- and sesquiterpenoids or almost completely sesquiterpenoids.The sesquiterpenoids are of greater interest and importance as far as chemotaxonomy and biological activity are concerned.

Table 2   Principal monoterpenes found in the volatile oils of Valeriana species

Species	Principal monoterpenes present (%)	Reference
V. alliarifolia	Borneol **2**	Bos *et al.* (1984)
	Bornyl acetate **4**	
V. dioica	Borneol **2**	Bos *et al.* (1984)
	Bornyl acetate **4**	
V. fauriei Briq.	Bornyl acetate **4** (50)	Rücker (1979)
	Camphene **5** (16)	
	α-pinene **6** (7)	
	β-pinene **7** (6)	
V. officinalis L.	Borneol **2**	Stoll *et al.*(1957)
	Bornyl isovalerate **3**	
	Bornyl acetate **4**	
	Camphene **5**	
	Camphor **10**	
	1-8 cineol **9**	
	Myrcene **8**	
	Camphene **12** (19)	Nikiforov *et al.* (1994)
	bornyl acetate **4** (13)	
	α-pinene **6** (10)	
	Bornyl isovalerate **3** (4)	
	Myrcene **8** (3)	
V. wallichii DC	Camphene **5**	Narayanan *et al.* (1964)
(syn. *V. jatamansii* Jones)	α-pinene **6**	
	β-cymene **11**	

Monoterpenes

In most samples of oil of *V. officinalis* examined, major constituents are borneol **2** and its isovaleric and acetyl esters **3, 4** (Stoll, 1957, Titz,1983) and the acetate was also a major compound detected in the headspace analysis of the stored roots (Nikiforov *et al.*, 1994).

Hydrocarbons such as camphene **5**, α-pinene **6**, β-pinene **7**, and myrcene **8** form a major part of the monoterpene component of some oils. Oxygenated compounds such as 1-8 cineol **9** and camphor **10** are found as minor components although they occur in significant amounts in the oil of some chemotypes of *V. officinalis* (Stoll *et al.*, 1957; Nikiforov *et al.*, 1994)

A summary of the major and important minor monoterpene constituents of volatile oils of *Valeriana* spp. found by various investigators is shown in Table 2.

Sesquiterpenes

The Valerianaceae is a source of a variety of sesquiterpenes including two unique ring systems. The sesquiterpenes are found as components of the volatile oils and in some species e.g. *V. edulis* Nutt ex Torr. et Gray, the volatile oil is comprised almost totally of sesquiterpenes (Hänsel *et al.*, 1994).

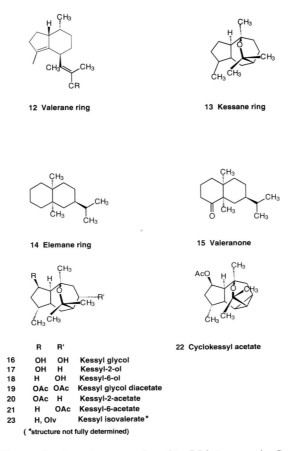

12 Valerane ring

13 Kessane ring

14 Elemane ring

15 Valeranone

22 Cyclokessyl acetate

	R	R'	
16	OH	OH	Kessyl glycol
17	OH	H	Kessyl-2-ol
18	H	OH	Kessyl-6-ol
19	OAc	OAc	Kessyl glycol diacetate
20	OAc	H	Kessyl-2-acetate
21	H	OAc	Kessyl-6-acetate
23	H, Olv		Kessyl isovalerate*

(*structure not fully determined)

Figure 3   Sesquiterpenes found in *Valeriana* species I

The two ring systems unique to the Valerianaceae are the valerenal type **12** and the kessane type **13**. Another important ring found in the family is the an elemane type **14** of which valeranone **15** is an important constituent. This compound was first isolated from oil of nard,derived from *Nardostachys jatamansii* DC.(a genus related to *Valeriana*). This oil has long been prized as a perfume in the Middle East and is mentioned in the Bible.

The kessane group of sesquiterpenes were the first to be characterised when Ukida (1944) isolated kessyl glycol **16** during an investigation of the oil from the roots of the Japanese plant *V. officinalis* var. *latifolia* Miq., now known as *V. fauriei* Briq. Subsequent studies resulted in the isolation of other alcohols **17,18** and corresponding acetates **19-21** from the same plant (Hikino, 1963). All the kessyl derivates exert some sedative activity and the diacetate **19** was shown to be the most potent of compounds **16-21**

	R	R'	
24	CHO	H	Valerenal *(Stoll et al., 1957)*
25	COOH	H	Valerenic acid *(Stoll et al., 1957)*
26	COOH	OH	Hydroxyvalerenic acid *(Buechi et al., 1960*
27	COOH	OAc	Acetoxyvalerenic acid *(Buechi et al., 1960*
28	CH₂OH	H	Valerenol *(Bos et al., 1986)*
29	CH₂OAc	H	E-Valerenyl acetate *(Bos et al., 1986)*
30	CH₂OIv	H	E-Valerenyl isovalerate *(Bos et al., 1986)*

31 $R = CH_2OOCC_4H_9$
    $R' = H$ Valerenyl *n*-valerate *(Bos et al., 1986)*

32 $R = CH_2OOCC_5H_{11}$
    $R' = H$ Valerenyl hexanoate *(Bos et al., 1986)*

33	R = Ac	Z-Valerenyl acetate *(Bos et al., 1986)*
34	R = Iv	Z-Valerenyl isovalerate *(Bos et al., 1986)*

Figure 4   Valerene derivatives found in *V. officinalis*

(Hikino, 1980). Recent studies have shown that α-kessyl alcohol (kessyl-2-ol) **17** kessanol (kessyl-6-ol) **18** and cyclokessyl actate **22** have antidepressant activity whereas the glycol **16** and its diacetate **19** do not possess this activity (Oshima *et al.*, 1995) The kessane sesquiterpenes have also been isolated from *V. officinalis* (Hazelhoff *et al.*, 1979) and *V.wallichii* , the latter including the novel isovalerate ester **23** (Bos *et al.*, 1992).

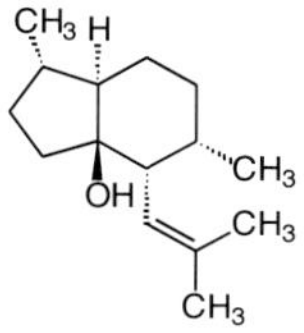

35 (-)-Pacifigorgiol

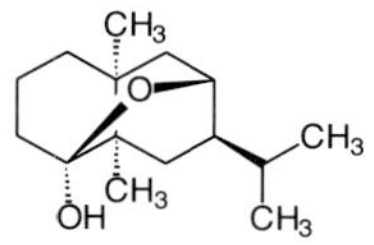

36 Cryptofaurinol

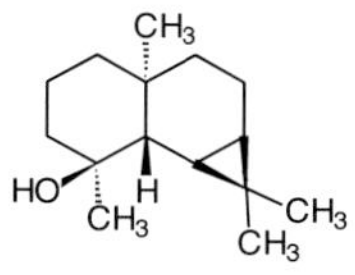

37 Maaliol

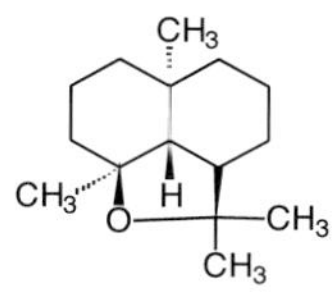

38 Maalienoxide

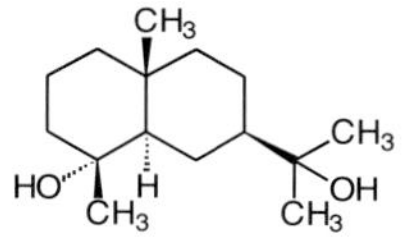

39 Cryptomeridiol

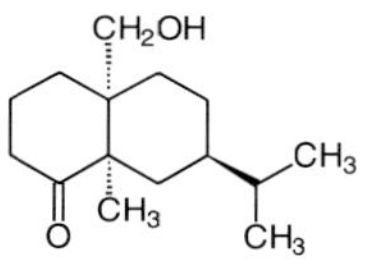

40 Kanokonol

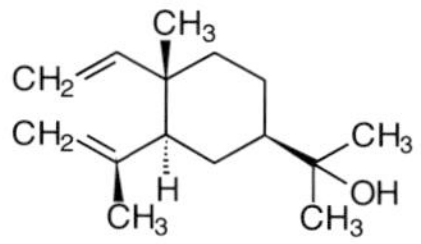

41 Elemol

Figure 5   Sesquiterpenes found in Valeriana species II

The investigation of the volatile oil of European *V. officinalis* by Stoll *et al.* (1957) resulted in the isolation of two sesquiterpenes having a new ring structure and these were named valerenal **24** and valerenic acid **25** respectively. Several representatives of this type of structure **24-34** have since been isolated from other *Valeriana* species (see Table 2) but valerenic acid **24** has so far not been found in any species other than *V. officinalis* and its presence is used as an identification test for *V. officinalis* in the European Pharmacopeia (1985). Valerenic acid **24** has also been shown to make a substantial contribution to the sedative and spasmolytic activity of the oil and extracts of *V. officinalis* (Hendriks *et al.*, 1981). The isolation of the related compound (-)-pacifigorgiol **35** (Bos *et al.*, 1986) is of interest as it is the optical isomer of a compound isolated from a coral.

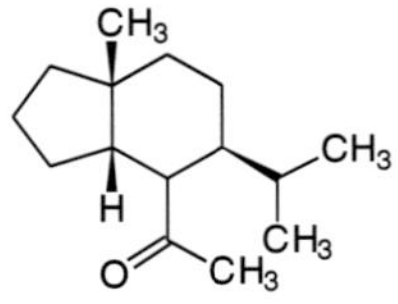

42 Faurinone

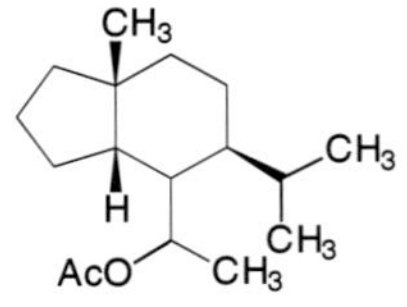

43 Fauronyl acetate

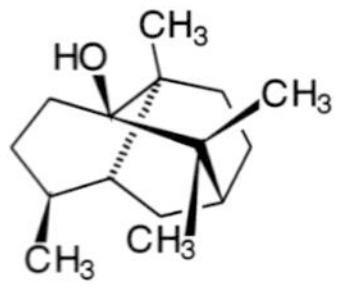

44   Patchouli alcohol

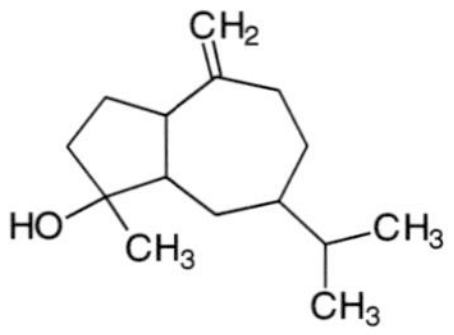

45   Nardol

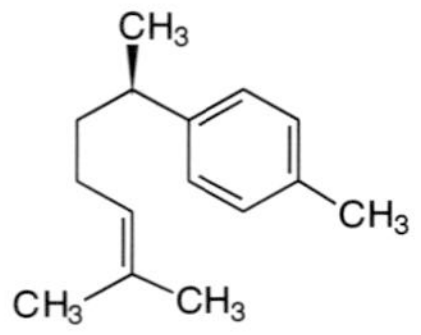

46 α- Curcumene

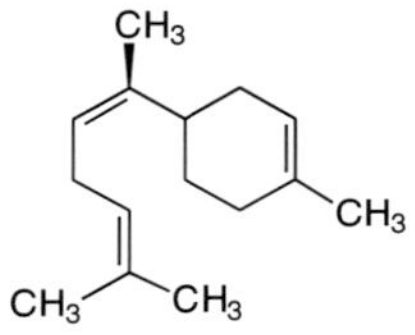

47   β–Bisabolene

Figure 6   Sesquiterpenes found in Valeriana species III

The third major group of sesquiterpenes found in Valeriana have an elemane ring structure and both hydrocarbons and oxygenated compounds are found, the latter type exhibiting the more pronounced pharmacological effects. Valeranone **15** was first isolated from *Nardostachys jatamansii* DC (Govindachari *et al.*, 1958) and has subsequently been detected in the oil of *V. officinalis* (Hendriks *et al.*, 1981), *V. fauriei* (Hikino *et al.*, 1968) and *V. wallichii* (Navayanan *et al.*, 1964). Related compounds are cryptofaurinol **36** found in *V. officinalis* , maaliol **37**, maalienoxide **38** and cryptomeridiol **39** from *V. wallichii* (Joshi *et al.*, 1968) and kanokonol **40** from *V. fauriei* (Oshima *et al.*, 1995). The open A ring compound elemol **41** is found in *V. officinalis* (Stoll *et al.*, 1957). Two related compounds faurinone **42** and fauronyl acetate **43** possess a 5-membered A ring instead of the 6-membered ring characteristic of the elemane type.

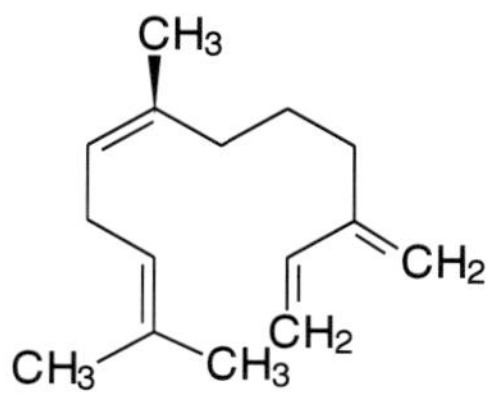

48 β–Farnesene 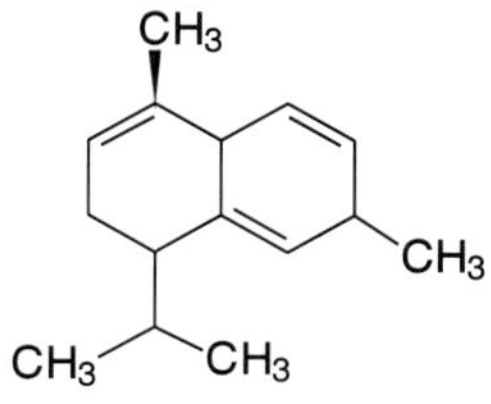   **49   1,4,9 Cadinatriene**

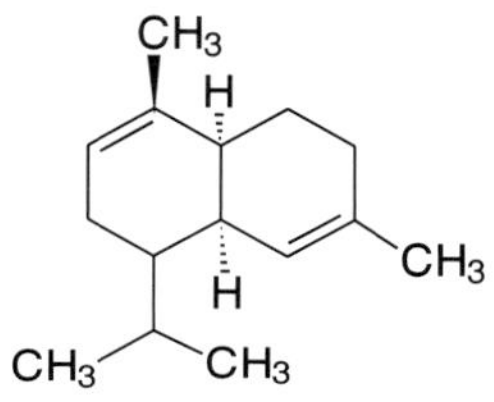

50 γ-Cadinene

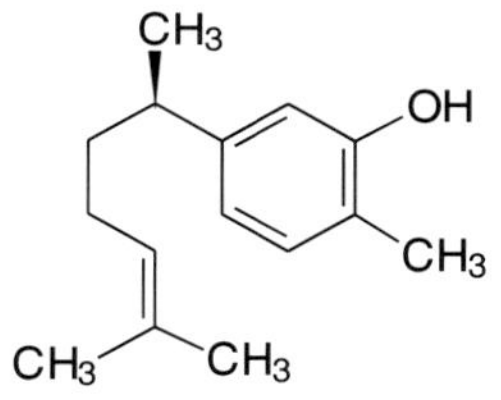

51   Xanthrorizol 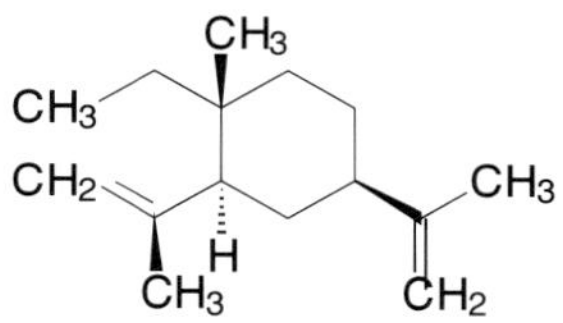   **52 β- Elemene**

Figure 7   Sesquiterpenes found in Valeriana species IV

Patchouli alcohol **44** has been found in the oil of *V. edulis, V. officinalis* and *V. wallichii* and in the last named species it forms the major component of the volatile oil.. The related compound nardol **45** has also been detected in *V. fauriei* (Hikino *et al.*, 1963).

In addition to these oxygenated compounds some hydrocarbon sesquiterpenes have been isolated from *Valeriana*. α-curcumene **46** and β-bisabolene **47** are derivatives of farnesene **48**, the precursor to all sesquiterpenes, and have been detected in *V. officinalis* and *V. wallichii*. The related compounds **49-51** are also found as minor constituents of the oils of some species (see Table 3). Hydrocarbons comprising other sesquiterpene ring systems are also found as constituents of volatile oils of *Valeriana*. and include β-elemene **52**, the patchoulenes **53, 54**, guajene **55** and α-copaene **56**.

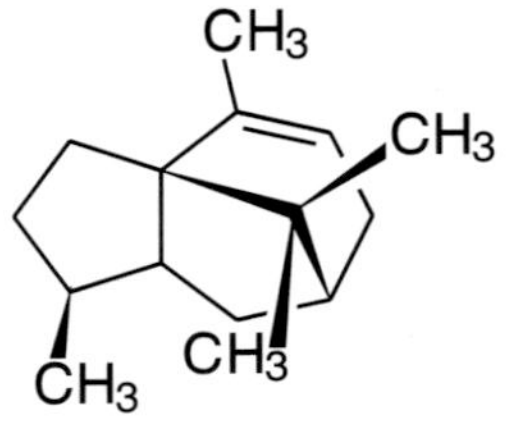

53   α-Patchoulene 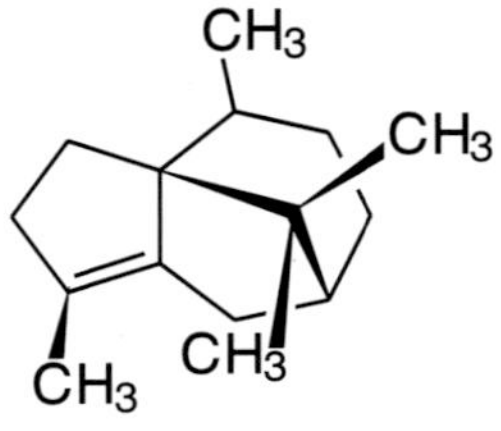   **54   β- Patchoulene**

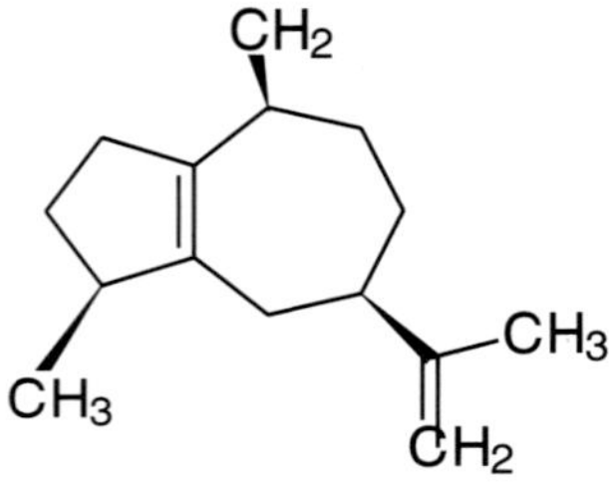

55   Guajene 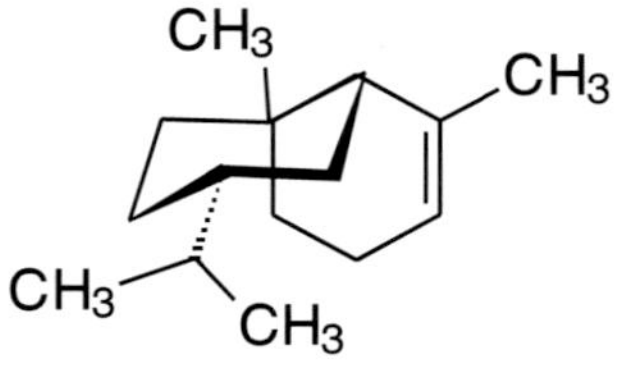   **56   α-Copaene**

Figure 8   Sesquiterpenes found in Valeriana species V

A summary of the sesquiterpenes found in *Valeriana* species is given in Table 3.

Variability in volatile oil composition

It is important to realise that the total amount, the constituents present and the proportion of each constituent in a volatile oil derived from a particular species varies considerably. This variation may be due to environmental factors such as climate and soil. Studies have shown that oil content in some plants varies greatly throughout the growing season and even on a diurnal basis and so variation may be due to the time of collection. The composition of the volatile oil also varies according to the part of the plant from which it is distilled but this is not of much concern with *Valeriana* since the total underground

Table 3   Major sesquiterpene components of Valeriana species investigated

V. alliarifolia	OXYGENATED COMPOUNDS Kessane **13**	Bos *et al.*, 1984
V. celtica ssp. norica	HYDROCARBONS α–Patchoulene **52** β–Patchoulene **53** OXYGENATED COMPOUNDS Patchouli alcohol **44** Valerenic acid **25** Valerenyl acetate **29**	Bicchi *et al.*, 1983
V. edulis	HYDROCARBONS 1,4,9-Cadinatriene **49** α–Copaene **56** β–Elemene **52** α–Guajene **55** α–Patchoulene **52** β–Patchoulene **53** OXYGENATED COMPOUNDS Patchouli alcohol **44**	Hendriks and Bos, 1984
V. fauriei	OXYGENATED COMPOUNDS Kessane **13** Faurinone **42** Kanokonol **40** Valeranone **15** Kessyl glycol **16** Kessyl-2-ol **17** Kessyl-6-ol **18** Kessyl diacetate **19** Kessyl -2-acetate **20** Kessyl-6-acetate **21** Cyclokessyl acetate **22**	Ukida (1944); Hikino *et al.*, (1963, 1980); Nishiya *et al.*, (1994, 1995), Oshima *et al.*, (1986, 1995)
V. officinalis	HYDROCARBONS β–Bisabolene **47** α–Curcumene **46** OXYGENATED COMPOUNDS Valerenic acid derivatives **24-34** (see Table) Faurinone **42** (-)-Pacifigorgiol **35** Cryptofaurinol **36** Valeranone **15** Patchouli alcohol **44** Kessyl alcohol **18**	Bos *et al.*, 1986a, 1986b; Hazelhoff *et al.*, 1979; Stoll *et al.*, 1957a
V. phu	OXYGENATED COMPOUNDS Patchouli alcohol **44** Valerenal **24** isomers	Bos *et al.*, 1984

Table 3   continued

V. wallichii DC	HYDROCARBONS
	α–Curcumene **46**
	β–Farnesene **48**
	α–Patchoulene **53**
	β–Patchoulene **54**
	OXYGENATED COMPOUNDS
	Cryptomeridiol **39**
	Kanokonol **40**
	Maaliol **37**
	Maalienoxide **38**
	Xanthrorizol **51**
	Patchouli alcohol **44**

Table 4   Yield of oil produced from underground organs of Valeriana species

Species	Yield of oil (%ov/w)	Reference
V. edulis	< 0.02	Hänsel *et al.*, 1994
V. fauriei	6.5–8.0	Rücker, 1979
V. officinalis	0.4 - 2.0	Bos *et al.*, 1984
V. wallichii DC	0.09–0.9	Bos *et al.*, 1992
V.wallichii DC(*V. jatamansii* Jones)	0.5–3.0	Hänsel *et al.*, 1994

parts are the only parts used and the oil occurs mainly in the endodermis cells of the roots. The yields of steam-distilled oil from various *Valeriana* species is shown in Table 4.

A major factor underlying the variation in volatile oil content, particularly the type of compounds present, is the genotype of the species which is used. Some studies have carried out in this respect and wild populations of *V. officinalis* in the Netherlands have been shown to comprise three chemical races i.e. phenotypes (Hazelhoff *et al.*, 1979). These were identified as Types A, B and C.

Type A contained no kessane derivatives, high amounts of valerenal **24** and moderate amounts of elemol **41** and valeranone **15**.

Type B contained no kessane derivatives or valeranone **15** but high amounts of elemol **41** and valerenal **24**.

Type C contained moderate amounts of kessane derivatives, elemol **41** and valerenal **24** together with high amounts of valeranone **15**.

Other terpenoids not present as volatile oils - the Valepotriates

The valepotriates

The valepotriates were first isolated in 1966 from *V. wallichii* (Thies, 1966) and from *Centranthus ruber* a member of a related genus of the Valerianaceae (Mannenstatter *et al.*, 1966). the isolation of these substances aroused considerable interest since they provided some answer to the discrepancy often observed between a high measured sedative/ tranquillising effect of Valerian extracts which could not be explained only on the basis of the amount of volatile oil present.

The term 'Valepotriate' arises from the part-acronym of **Val**eriana **epo**xy **tri**ester. The monoterpene skeleton of the valepotriates is essentially the same as that of the group of monoterpene glycosides known as iridoids, a group of compounds found widely throughout the more highly-evolved dicotyledons, but the valepotriates are unusual since, in most cases, no sugar residue is attached and also one or more ester side chains are present. The hydrolysis of these ester bonds and release of the free acids on storage makes a large contribution to the characteristic odour of dried *Valeriana* species.

The first three valepotriates isolated were obtained from *V. wallichii* and were named valtrate **57**, acevaltrate **58** and didrovaltrate **59** (Thies, 1966). The 'didro' series lacks the 5-6 double bond present in the corresponding compounds. The first valepotriates isolated all contained an 8-epoxy group and this has been shown to possess alkylating properties similar to epichlorhydrin and NN-dimethyl-N-(2-chloroethyl)-amine which result in *in vitro* cytotoxicity. (Braun *et al.*, 1982). No such toxicity could,however, be detected in *in vivo* studies, even at very high doses (Tortarolo, 1982).

Several members of the two original types of compounds isolated have now been reported **57–76** and differ in the type and position of the ester side chains.

Concern about the alkylating property of the epoxide group prompted the search for valepotriates where it was absent and several types of such compounds **77-88** have now been isolated (Hölzl *et al.*, 1976; Finner *et al.*, 1984; Koch and Hölzl, 1985). The largest group of these are known as valtrate hydrines **80-88**.

The valepotriates first isolated contained no sugars but valepotriate glycosides **89-93** with the sugar linked through C-1, as found in most iridoids, or through an alcohol residue at C-4, have since been isolated from *V.officinalis* and *V. wallichii* (Thies, 1970; Taguchi and Endo, 1974; Endo and Taguchi, 1977).

A valepotriate containing a chlorine atom, valechlorine **94**, was isolated from *V. officinalis* (Popov *et al.*, 1973) and this is noteworthy since a halogen incorporated into a secondary metabolite is unusual amongst the flowering plants.

A standard mixture stated to consist of   80% didrovaltrate,15% valtrate and 5% acevaltrate is marketed in Germany under the name Valmane and is used extensively as a mild to moderate sedative. Recently sophisticated NMR studies have shown that the tablets consist of a mixture of six compounds. The presence of valtrate and didrovaltrate was confirmed but the acevaltrate was shown to be a mixture of 1-α-acevaltrate and 7-β-acevaltrate whilst the remaining two compounds were novel and had not previously been reported from any plant (Lin *et al.*, 1991).

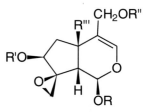

	R	**R'**	**R"**	
57	Iv	Iv	Ac	Valtrate *(Thies, 1968a)*
58	Aiv	Iv	Ac	1-α-Acevaltrate *(Thies, 1968a)*
60	Iv	Bav	Ac	7-β-acetvaltrate *(Lin et al., 1991)*
61	Iv	Ac	Iv	Isovaltrate *(Thies et al., 1973)*
62	Iv	Ac	Ac	Diavaltrate *(Marekov et al., 1983)*
63	Miv	Iv	Ac	Homovaltrate 1 *(Thies et al., 1973)*
64	Iv	Ac	Miv	Homovaltrate 2 *(Thies et al., 1973)*
65	Iv	Ac	Aav	11-Acevaltrate *(Thies, 1971)*
66	Iv	Aav	Ac	Homoacevaltrate *(Thies, 1981)*
67	Iv	Hiv	Ac	Hydroxyvaltrate *(Marekov et al., 1983)*
68	Iv	Aic	Ac	Isohomoacevaltrate *(Marekov et al., 1983)*
69	Cr	Iv	Ac	Seneciovaltrate *(Koch and Hölzl, 1985)*
70	Iv	Miv	Hiv	*(Lin et al., 1991)*
71	Iv	Iv	Hiv	*(Lin et al., 1991)*
72	Iv	H	Iv	Deacetylisovaltrate *(Popov, 1974)*

	R	**R'**	**R"**	**R'''**	
59	Iv	Ac	Iv	H	Didrovaltrate *(Thies, 1968a)*
73	Iv	Ac	Iiv	OH	IVHD valtrate *(Stahl and Schild, 1969)*
74	Iv	Iv	Ac	H	Isodidrovaltrate *(Kucaba et al., 1980)*
75	Miv	Ac	Iv	H	Homodidrovaltrate *(Thies, 1968a)*
76	Ac	Iv	Aav	H	AHD valtrate *(Marekov et al., 1983)*

Figure 9   Iridoids found in Valeriana species I

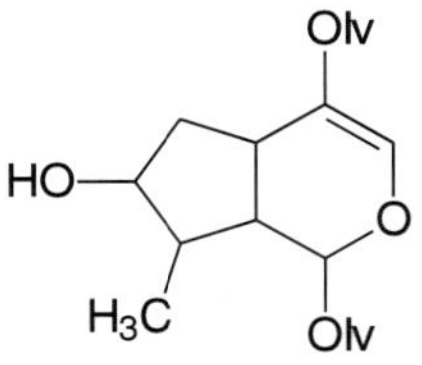

77 Nardostachin

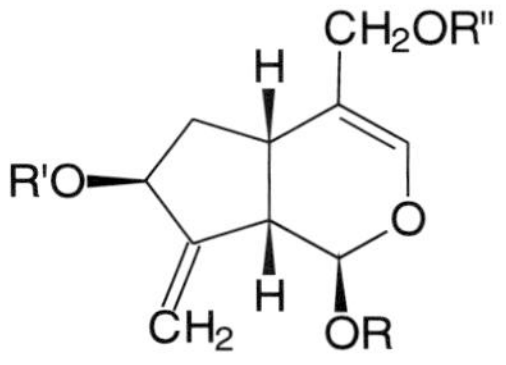

	R	R'	R"	
78	Iv	Ac	Iv	**8,11-Desoxidodidrovaltrate** *(Thies, 1968a)*
79	Miv	Ac	Iv	**8,11-Desoxidohomodidrovaltrate** *(Thies, 1968a)*

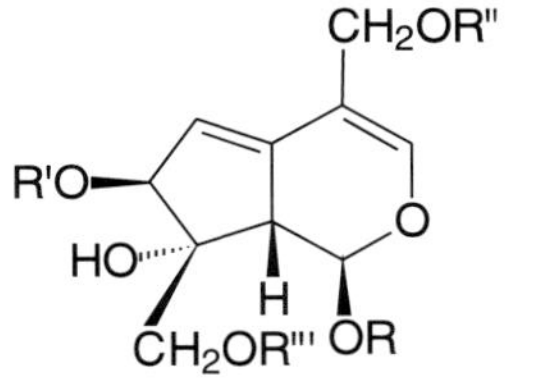

	R	R'	R"	R"'		
80	Iv	Iv	Iv	Ac	**Valtrate hydrine B1**	*(Hölzl et al., 1976)*
81	Iv	Iv	Ac	Ac	**Valtrate hydrine B2**	*(Hölzl et al., 1976)*
82	Biv	Iv	Iv	Ac	**Valtrate hydrine B3**	*(Hölzl et al., 1976)*
83	Iv	Ac	Iv	Ac		*(Finner et al., 1984)*
84	Iv	Iv	Ac	Iv	**Valtrate hydrine B4**	*(Koch and Hölzl, 1985)*
85	Iv	Iv/Miv	Ac		**Valtrate hydrine B5**	*(Koch and Hölzl, 1985)*
86		Iv/Miv	Ac	Ac	**Valtrate hydrine B6**	*(Koch and Hölzl, 1985)*
87	Cr	Iv	Iv	Ac	**Valtrate hydrine B7**	*(Koch and Hölzl, 1985)*
88	Aav	Iv	Iv	Ac	**Valtrate hydrine B8**	*(Koch and Hölzl, 1985)*

Figure 10   Iridoids found in Valeriana species II

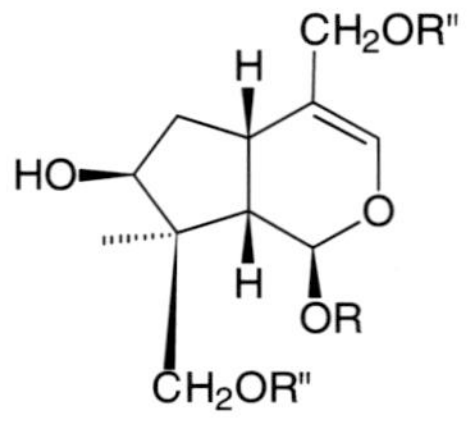

	R	R'	R"		
89	Glucose	Iv	H	Valerisodatum	*(Thies et al., 1970)*
90	Iv	Glucose	OH	Kanokoside B	*(Endo and Taguchi, 1977)*
91	Iv	Gentiobiose	OH	Kanokoside D	*(Endo and Taguchi, 1977)*

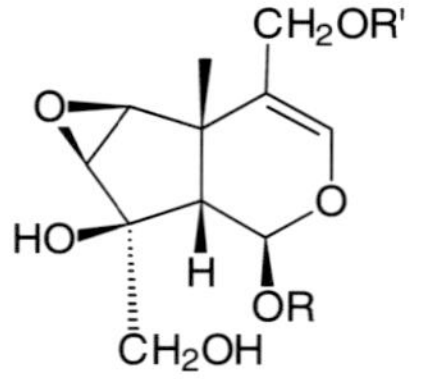

	R	R'		
92	Iv	Glucose	Kanokoside A	*(Endo and Taguchi, 1977)*
93	Iv	Gentobiose	Kanokoside B	*(Endo and Taguchi, 1977)*

Figure 11   Iridoids found in Valeriana species III

It has been shown that the composition of valepotriates in *V. wallichii* varies according to the chemical race (Iglesias and Vila, 1985). Two races exist, one of which has a mainly valtrate present and the other mainly didrovaltrate.

A summary of the valepotriates reported from each species of *Valeriana* so far investigated is given in Table 5.

Compounds produced by tissue culture

The commercial interest in *Valeriana* species as a source of valepotriates has prompted investigation into the use of cell cultures as a means of production. Studies on valepotriate

Table 5 Valepotriates found in Valeriana species

Species	Valepotriates present	Reference
V. alliarifolia roots	Valtrate hydrines B1, B2 **80-82** Valtrate hydrines B3-B8 **83-88**	Hölzl and Koch (1984) Koch and Hölzl (1985)
V. cardamines	Valtrate **57** Acevaltrate **58**	Fursa *et al.* (1884)
V. edulis spp. *procera* roots	Valtrate **57** Acevaltrate **58** Isovaltrate **61** Didrovaltrate **59** IVHD valtrate **73**	Tittel *et al.* (1978) Denee *et al.* (1979)
V. fauriei	Valtrate **57** Acevaltrate **58** IVHD-valtrate **73** Kanokosides A-D **90-93**	Stahl and Schild (1971) Endo and Taguchi (1977)
V. kilimandascharica leaf	Valtrate **57** Isovaltrate **61** Acevaltrate**58** Didrovaltrate **59** IVHD valtrate**73** (5.89)	Dossaji and Becker (1981)
stem	Valtrate **57** Isovaltrate **61** Didrovaltrate **59** (3.17)	Dossaji and Becker (1981)
flower	Valtrate **57** Isovaltrate **61** Didrovaltrate **59** (3.93)	Dossaji and Becker (1981)
rhizome	Valtrate **57** Isovaltrate Acevaltrate **58** Didrovaltrate **59** IVHD valtrate**73** (5.15)	Dossaji and Becker (1981)
V. microphylla	Acevaltrate **58** Diavaltrate **62** Isovaltrate **61** Nardostachin **77**	Bach *et al.* (1993)
V. officinalis	Valtrate **57** Acevaltrate **58** Isovaltrate **61** Deacetylisovaltrate **72** Didrovaltrate **59** IVHD-valtrate **73** Valerisodatum **88** Valechlorine **94** Kanokosides A, C, D **91-93**	Popov *et al.* (1974) Titz *et al.* (1982)

Table 5   continued

Species	Valepotriates present	Reference
V. sitchensis spp. *scouleri*		Forster *et al.* (1984)
V. spryginii	Valtrate **57** Acevaltrate **58**	Zaitsev *et al.* (1985)
V. thalictroides	(14.5%)	Becker *et al.* (1983)
V. tiliaefolia	Valtrate hydrines B1-B3 **80-82**	Hölzl *et al.* (1976)
V. wallichii leaves *V. wallichii* roots	Hölzl and Jurcic, K (1975) Valtrate **57** Acevaltrate **58** Isovaltrate **61** Didrovaltrate **59** IVHD valtrate **73** (5%) Valerosidatum **88** (5%)	Thies and Funke (1966a, 1966b)
V. vaginata	Isodidrovaltrate **74**	Kucaba *et al.* (1980)

production in callus and root differentiated tissue of *V. officinalis* showed that levels of valepotriates were formed at least equivalent to those occurring naturally in the root differentiated cultures (Violon et al, 1984)

Cell suspension cultures of *V. wallichii* were found to produce the valepotriates found in the parent plant and also six compounds not previously reported (Table 6). Most of these were of the diene type. Treatment of the culture with colchicine resulted in a higher yield of valepotriates (Becker and Chavadej, 1985).

Hairy root cultures of *V. officinalis* var. *sambucifolia* have been shown to produce isovaltrate and IVHD valtrate (Granicher *et al.*, 1992)

Decomposition products and metabolites

The valepotriates hydrolyse quite rapidly and are not present in significant amounts in aqueous or dilute alcoholic extracts after a few days. They are also metabolised in the gastro-intestinal tract to yield the breakdown products baldrinal **95**, homobaldrinal **96,** deacybaldrinal **97** and valtroxal **98** which consist of an unsaturated version of the ring skeleton (Thies, 1968; Schneider and Willems, 1979; Schneider and Willems 1982). There is evidence that these compounds are partly responsible for the sedative activity of some *Valeriana* samples since they are well-absorbed from the gut and have been shown to significantly decrease motility of mice (Schneider and Willems, 1982). Homobaldrinal **96** and valtroxal **98** have been shown to possess greater sedative activity than baldrinal **95** and deacylbaldrinal **97** (Wagner 1980; Schneider and Willems, 1982).

Table 6   Compounds detected in cell suspension cultures of V. wallichii
(Becker and Chavadej, 1985)

Cyclodienes

Known compounds

	R	R'	R"
Valtrate 57	Iv	Iv	Ac
1-α–Acevaltrate 58	Aiv	Iv	Ac
Homovaltrate 63	Miv	Iv	Ac
Isovaltrate 61	Iv	Ac	Iv
Diavaltrate 62	Iv	Ac	Ac

Previously unknown compounds subsequently detected in plant material

	R	R'	R"
	Miv	Ac	Iv
	Miv	Miv	Ac
	Aav	Iv	Ac
	Iv	Ac	Ac
	Iv	Miv	Hiv
	Iv	Iv	Hiv

Compounds produced only by tissue culture

	R	R'	R"
	Aav	Bav	Ac
	Aav	Miv	Ac
	Aav	Bav	Ac
	Iv	Ac	Hiv
	Miv	Ac	Ac

Known compounds

	R	R'	R"	R'"
Didrovaltrate 59	Iv	Ac	Iv	H
IVHD-Valtrate 72	Iv	Ac	Iiv	OH

Compound produced only by tissue culture

	R	R'	R"	R'"
	Iv	Ac	Iv	OH

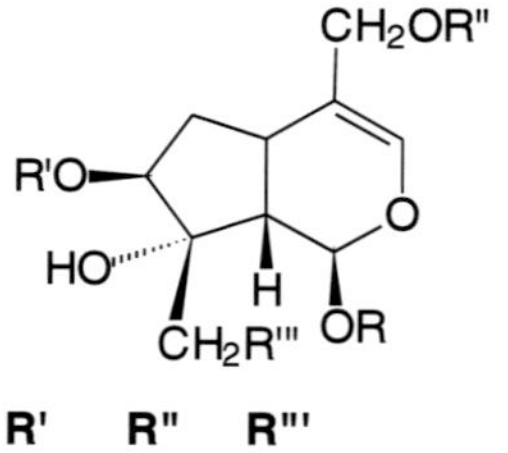

	R	R'	R"	R"'	
94	Iv	Iv	Ac	Cl	**Valechlorine** *(Popov et al., 1973)*

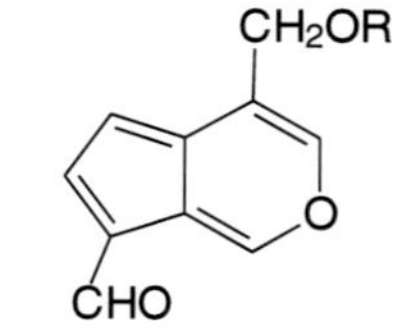

	R	
95	Ac	**Baldrinal** *(Thies, 1968)*
96	Iv	**Homobaldrinal** *(Thies, 1968)*
97	H	**Deacylbaldrinal** *(Schneider and Willens, 1979)*

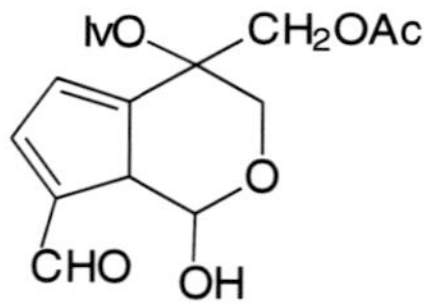

98	**Valtroxal** *(Thies, 1968a)*

Figure 12   Alkaloids found in Valeriana species

NITROGENOUS COMPOUNDS

Alkaloids

In some medicinal plants alkaloids are the most important secondary metabolites but this is not the case with *Valeriana* although alkaloids are present in small amounts. The presence of alkaloids in this genus has been known since the end of the nineteenth century (Walliczewski, 1891) but it was not until 1967 that two alkaloids **99** and **100** were isolated from *V. officinalis* (Torsell and Wahlberg, 1967). Two further related alkaloids valeranine **101** and actinidine **102** have also been isolated from the same species (Francke *et al.*, 1970; Johnson and Waller, 1971; Buckova *et al.*, 1977; Janot *et al.*, 1979) The structural similarity between these alkaloids and the valepotriates is noticeable and it is possible that they may be artefacts since it is well-known that iridoids can incorporate a

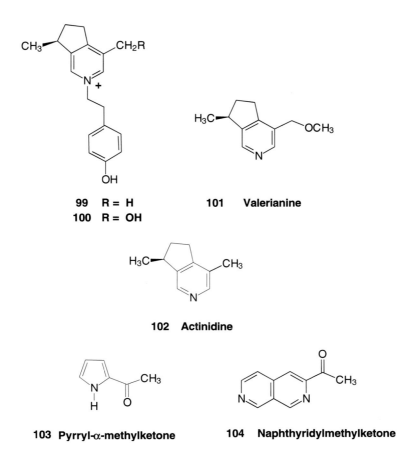

Figure 13 Alkaloids found in Valeriana species

nitrogen atom in the presence of ammonia, a reagent commonly used in extraction of alkaloids. The reports of the isolation of the first two compounds does not state the nature of the basifying agent so a conclusion cannot be reached as to whether they are artefacts or not. Valeranine has also been reported from *V. wallichii* (Bos *et al.*, 1992).

Another type of alkaloidal compound pyrryl -α-methylketone **103** and naphthyridyl-methylketone **104** have also been isolated by Cionga (1936) and Janot and his co-workers (1979).

All of these compounds are present in small amounts and it is very unlikely that they make a major contribution to the activity of total extracts. **99** has shown choline esterase inhibitory properties *in vitro* but these were not demonstrated *in vivo* in rabbits (Torsell and Wahlberg, 1967). Actinidine **102** exhibits antibiotic properties (Buckova et al., 1977; Johnson and Waller 1971) and the methylketone derivatives have been shown to have a sedative effect (Cionga, 1936).

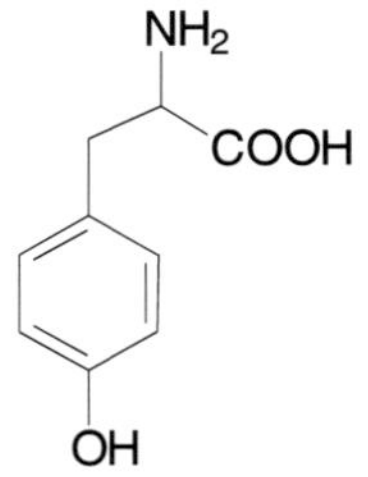

105 Tyrosine

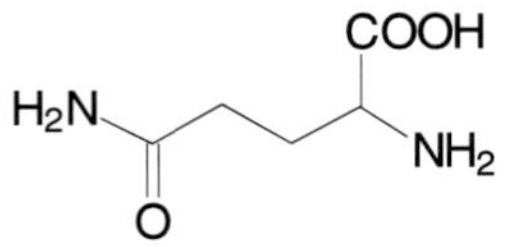

106 Glutamine

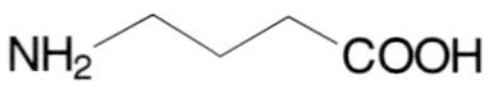

107 GABA

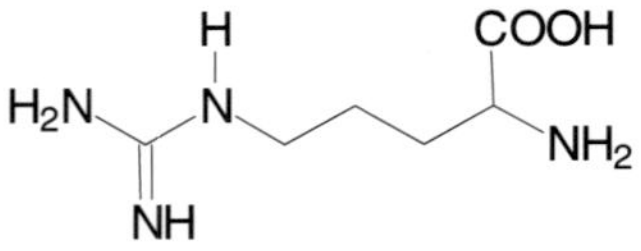

108 Arginine

Figure 14 Amino acids found in *Valeriana* species

Amino acids

V. officinalis contains appreciable amounts of amino acids. In one study tyrosine **105,** glutamine **106** and GABA (γ-aminobutyric acid) **107** were found to be the major compounds present (Hänsel and Schulz, 1981). More recent work by Santos *et al.*, 1994) demonstrated the presence of high levels of glutamine **106** (13.4mM) and arginine **108** (25.5mM) with a relatively high concentration of GABA **107** (4.56mM) in aqueous extracts of *V. officinalis.*

The quantities of GABA and glutamine were considered sufficient to account for the release of radiolabelled GABA from synaptosomes which was observed when they were treated with Valerian extract.

PHENYLPROPANOIDS

The phenylpropanoids constitute important secondary metabolites found in the flowering plants and are formed by the shikimic acid pathway. A characteristic of compounds formed by this route is the presence of one or more C6(aromatic ring)-C3 units in the molecule. In *Valeriana* four types of phenylpropanoids are found as relatively minor constituents which have not been demonstrated to play a significant role in any pharmacological effect.

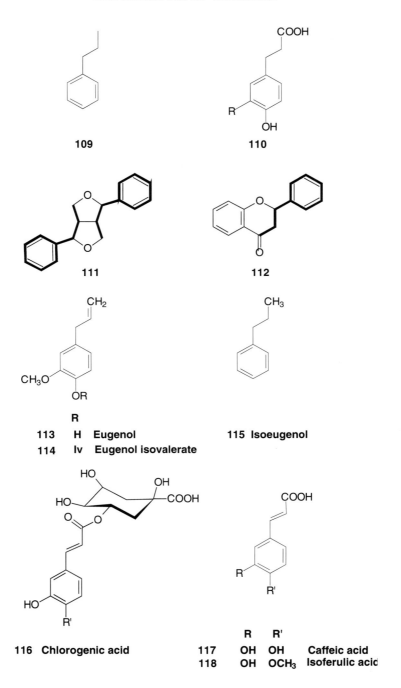

Figure 15   Phenylpropanoids of *Valeriana*

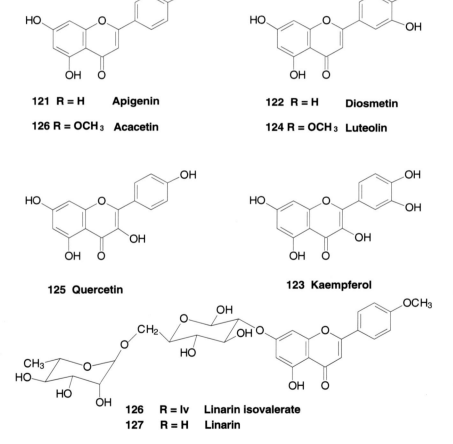

119   R = OH   (+)-1-Hydroxypinoresinol
120   R = H   Pinoresinol

121 R = H   Apigenin   **122 R = H**   Diosmetin

126 R = OCH₃ Acacetin   **124 R = OCH₃** Luteolin

125 Quercetin   **123 Kaempferol**

126   R = Iv   Linarin isovalerate
127   R = H   Linarin

Figure 16   Lignans and flavonoids of *Valeriana*

The four types are:-

a)   the volatile oil constituents consisting solely of one unit **109**
b)   the phenolic acids derived from one unit **110**
c)   the lignans **111** consisting of two units joined through the C3 portions of the molecule
d)   the flavonoids **112** formed by condensation of the phenylpropanoid unit with an aromatic ring derived from the conjunction of three acetates.

Volatile oil constituents

The phenylpropanoids form only a small part of the volatile oil of *Valeriana*. Eugenol **113** and its isovaleryl ester **114** as well as the isovalerate of isoeugenol **115** have been detected in *V. fauriei* and *V. officinalis* (Hendriks *et al.*, 1981)

Phenolic acids

The phenolic acids derived from the shikimate pathway are probably present in most plant species to some extent either as free compounds or combined with sugars. Chlorogenic acid **116** and caffeic acid **117** were reported from *V. officinalis* a considerable time ago (Fichter, 1939) and more recently isoferulic acid **118** has been reported (Stoll and Seebeck, 1957; Rybal'chenko *et al.*, 1976).

Lignans

The first and only report of lignans in Valeriana appeared in 1993 with the isolation of (+)-hydroxypinoresinol **119** as the major constituent and pinoresinol **120** from *V. microphylla* (Bach *et al.*, 1993).

Flavonoids

Flavonoids are present in nearly all flowering plant species and represent a large number of compounds. The flavonoids of many species of Valeriana, in both aglycone and glycosidal forms have been investigated by workers in the former Soviet Union and the aglycones found are listed in Table 7. *V. officinalis* have been investigated by Rybal'chenko and co-workers who isolated luteolin **124**, diosmetin **122** and kaemferol **123** (Rybal'chenko *et al.*, 1976). An unusual compound, linarin isovalerate **126** was isolated from *V. wallichii* together with linarin **127** by Thies (1968b).

FATTY ACIDS AND THEIR ESTERS

The fatty acids are primary metabolites rather than secondary metabolites but some are of interest because of their role as essential constituents in the diet and as precursors in biochemical processes involved with inflammatory conditions and several other disease states. The seeds of *V. officinalis* produce an oil rich in unsaturated fatty acids such as linoleic acid **128** (Dolya, 1983). The methyl ester of eicosanoic acid **129** has recently been isolated and quantified in *V. wallichii* (Pande *et al.*, 1994).

Table 7   Flavonoid aglycones found in Valeriana species

Valeriana *species*	*Flavonoid aglycones present*	*Reference*
V. cardamines	Apigenin **121** Diosmetin **122** Kaempferol **123** Luteolin **124** Quercetin **125**	Fursa et al (1984)
V. chionophila	Acacetin **126** Apigenin **121** Diosmetin **122** Luteolin **124** Quercetin **125**	Trzhetsinskii et al (1982)
V. eriophylla	Apigenin **121** Diosmetin **122** Kaempferol **123** Luteolin **124** Quercetin **125**	Fursa et al (1984)
V. exaltata	Apigenin **121** Diosmetin **122** Kaempferol **123** Luteolin **124** Quercetin **125**	Rybal'chenko et al (1976)
V. fedtschenkoi	Acacetin **126** Apigenin **121** Diosmetin **122** Kaempferol **123** Luteolin **124** Quercetin **125**	Trzhetsinskii et al (1982)
V. ficariifolii	Acacetin **126** Apigenin **121** Diosmetin **122** Kaempferol **123** Luteolin **124** Quercetin **125**	Trzhetsinskii et al (1982)
V. nitida	Apigenin **121** Diosmetin **122** Kaempferol **123** Luteolin **124** Quercetin **125**	Rybal'chenko et al (1976)
V. officinalis	Diosmetin **122** Kaempferol **123** Luteolin **124**	Rybal'chenko et al (1976)

Table 7   continued

Valeriana *species*	*Flavonoid aglycones present*	*Reference*
V.palustris	Apigenin **121** Diosmetin **122** Kaempferol **123** Luteolin **124** Quercetin **125**	Fursa (1983)
V.spryginii	Apigenin **121** Luteolin **124**	Zaitsev et al (1985)
V.tuberosa	Acacetin **126** Apigenin **121** Diosmetin **122** Kaempferol **123** Luteolin **124** Quercetin **125**	Fursa et al, (1983)
V.turkestanica	Acacetin **126** Apigenin **121** Diosmetin **122** Kaempferol **123** Luteolin Quercetin **125**	Fursa et al (1981)

128 Linoleic acid

129 Methyl eicosanoate

Figure 17   Fatty acid derivatives found in Valeriana

REFERENCES

Bach, K.K., Ghia, F. and Torssell, K.B.G. (1993) Valtrates and Lignans in *Valeriana microphylla*. *Planta Medica* **59**,478–479.

Balandrin, M.F., Vanwagenen, B.C. and Cordell, G.A. (1994) Ammonia-mediated degradation of valepotriates from *Valeriana* species. *Abstracts of papers of the American Chemical Society* **208**, 73–AGFD

Becker, H. and Chavadej, S. (1985) Valepotriate Production of Normal and Colchicine-Treated Cell Suspension Cultures of *Valeriana wallichii*. *J.Nat. Prod.* **48**, 17–21.

Bicchi, C., Sanda, P., Schelfaut, M. and verzela, M. (1983) Studies on the essential oil of *Valeriana celtica* L. *J. High Resolut. Chromatogr. Chromatogr. Commun.* **6**, 213–215.

Bos, R;, Hendriks, H. Kloosterman, J. and Sipma, G.(1983)A structure of faurinone, a sesquiterpene isolated from Valeriana officinalis. Phytochemistry **22**, 1505–1506.

Bos, R., Hendriks, H. and Mastewnbroek, C. (1984) Terpenoide Inhaltstoffe von *Valeriana* species. *Oesterr. Apoth. Ztg* **38**, 49.

Bos, R., Hendriks, H., Bruins, A.P., Kloosterman, J. and Sipma, G.(1986a) Isolation and Identification of Valerenane Sesquiterpenoids from *Valeriana officinalis*. Phytochemistry **25**, 133–135.

Bos, R., Hendriks, H., Kloosterman, J. and Sipma, G.(1986b) Isolation of the Sesquiterpene Alcohol (-)-Pacifigorgiol from Valeriana officinalis. *Phytochemistry* **25**, 1234–1235.

Bos, R., Wordenbag, H.J. and Scheffer, J.J.C. (1993) Seasonal variation of the essential oil, valerenic acid derivatives and valepotriates in Valeriana officinalis roots. Planta Med. **59**, A698.

Buchbauer, G., Jager, W., Jirovetz, L., Meyer, F. and Dietrich, H. (1992) Wirkungen von Baldrianol, Borneol, Isoborneol, Bornylacetate und Isobornyl acetate auf die Motilitat von Versuchstieren (Mausen) nach Inhalation. *Pharmazie* **47**, 620–622.

Buckova, A. (1977) Active principles in *Valeriana officinalis* L. *Cesk. Far.* **26**, 308–309.

Buechi, G., Popper, T.L., and Staufacher D. (1960) Terpenes XIV. The structure of valerenic acid. *J.Amer. Chem. Soc.* **82**, 2962.

Chari, V.M., Jordan, M., Wagner, H. and Thies, P.W. (1977) A ^{13}C–NMR study of the structure of an acyl-linarin from Valeriana wallichii. Phytochemistry **16**, 1110–1112.

Cionga, E. (1936) Gegenwart von Pyrryl-α-methylketon in stabilisierter Valeriana officinalis. *C.R. hebd. Seances Acad. Sci.* **200**. (CA **29**, 3770).

Dolya, V.S. (1983) The effect of irrigation on the contents and chemical composition of fatty acids in seeds of some cultivated species of *Valeriana* L. *Rastit. Resur.* **22**, 348–51. (CA **105**, 168939)

Dossaji, S.F. and Becker, H. (1981) HPLC separation and quantitative determination of Valepotriates from *Valeriana kilimandascharica Planta Medica* **43**, 179–182.

Endo, T. and Taguchi, H., (1977) The Constituents of Valerian Root. The Structures of four New Iridoid Glycosides, Kanakoside A, B, C and D from the Root of 'Hokkaikisso'. *Chem. Pharm. Bull.* **25**, 2140–2142.

European Pharmacopeia 2nd Edn. (1985) Part II–9 p. 453 Maissonneuve, St ruffine, France.

Fichter, M. (1939) Isolierung von Chlorogensäure aus Baldrian *Pharm. Act. Helvet.* **14**, 163–170.

Finner, E., David, S. and Thies, P.W. (1984) Über die Wirkstoffe des Baldrians XV. Konstitutive Zuordnung der Acyloxy-Substituenten in Valepotriathydrinen vom Dientyp via ^{13}C NMR Spektroskopie. Planta Med. **50**, 4–6.

Förster W, Becker, H. and Rodriguez, E. (1984) HPLC analysis of valepotriates in the North American genera Plectritis and Valeriana. *Planta Medica* **50**, 7–9.

Francke, B., Petersen, U. and Hüper, F. (1970) Valerianine, a tertiary monoterpene alkaloid from Valerian. *Angewandte Chemie* (International Edition) **9**, 891.

Fursa, N.S., Belyaeva, L.E and Rybal'chenko, A.S. (1981) Phenolic compounds of the aboveground parts of Valeriana III Composition of Valeriana turkestanica flavonoids. Khim. Proer. Soedin. 98 (**CA 95**, 21263).

Fursa, N.S.(1983) Valeriana palustris Kreyer flavonoids. Rastit. Resur. **19**, 216–218 (**CA 99**, 3068).

Fursa, N.S. and Belyaeva, L.E. (1983) Flavonoid composition of vegetative and reproductive organs of Valeriana tuberosa L. Ukr. Bot. Zh. **40**, 36–38.

Fursa, N.S.,Trzhetsinskii, S.D., Zaitsev, V.G. and Gorbunov, Y.N.(1984) Phenolic compounds of the aboveground part and iridoids of the subsurface part of Valeriana VI. Flavonoids and vlepotriates of Valeriana eriophylla and Valeriana cardamines. Khim. Proer. Soedin. 249 (**CA 101**:51740).

Govindachari, T.R., Rajadurai, S. and Pai, B.R. (1958) Struktur von Jatamanson. *Chemische Berrichte* **91**, 908–910.

Granicher, F., Christen, P. and Kapetanidis, I. (1992) Production of valepotriates by hairy root cultures of *Valeriana officinalis* var. *sambucifolia Planta Medica* **58**, A614.

Granicher, F., Christen, P., Kamalaprija, P. and Burger, U. (1995) An iridoid diester from *Valeriana officinalis* var. *sambucifolia* hairy roots. *Phytochemistry* **38**, 103–105.

Grote (1832) Aus der wurzel von Valeriana officinalis vogl; A **4**, 229.

Hänsel, R., Keller, K., Rimpler, H. and Schneider, G. (eds) (1994) *Hager's Handbuch der Pharmazeutischen Praxis* 5th edn. **6**, 1071–1095, Springer, Berlin.

Hänsel, R. and Schultz, J. (1982a) Valerensäuren und Valerenal als Leitstoffe des offinizellen Baldrians. *Deutsche Apotheker Zeitung* **122**, 215–219.

Hänsel, R. and Schultz, J. (1982b) GABA and other amino acids in Valerian root. *Arch. Pharm.* **314**, 380–381.

Hazelhoff, B., Smith, D., Malingré, T.M. and Hendriks, H. (1979) The essential Oil of Valeriana officinalis. L. *Pharmaceutisch Weekblad* **114**, 443–449.

Hendriks, H., Bos, R., Allersma, D.P., Malingré, T.M. and Koster, A.S. (1981) Pharmacological Screening of Valerenal and some other Components of the essential Oil of *Valeriana officinalis. Planta Medica* **42**, 62–68.

Hendriks, H., Smith, D. and Hazelhoff, B. (1977) Eugenyl isovalerate and isoeugenyl isovalerate in the essential oil of Valerian root. Phytochemistry 16, 1853–1854.

Hikino, H., Hikino, Y., Koto, H., Takeshuta, Y. and Takemoto, T. (1963) Constituents of Kesso root. *Yagugaku Zasshi* **83**, 219–230.

Hikino, H., Hikino, Y., Agatsuma, K. and Takemoto,T. (1968) Structure and absolute configuration of faurinone. Chem. Pharm. Bull. **16**, 1779–1783.

Hikino, H., Hikino, Y., Kobinata, H., Aizawa, A., Konno, C. and Ohizumi, Y. (1980) Study on the efficacy of oriental drugs 18: Sedative properties of Valeriana roots. *Shoyakugaku Zasshi* **34**, 19–24.

Hölzl, J., Chari, V.M., and Seligmann, O. (1976) Zur Struktur von drei Genuinen Valtrathydrinen aus Valeriana tiliaefolia. *Tetrahedron Letters* 1171–1174.

Hölzl, J. and Jurcic, K (1975) Valepotriates in the leaves of *Valeriana jatamansii. Planta Medica* **27**, 133–139.

Houghton, P.J., (1988) The Biological Activity of Valerian and Related Plants, *Journal of Ethnopharmacology* **22**, 121–142.

Iglesias J. and Vila, R. (1985) Study on valepotriate levels in cultivated *Valeriana wallichii. Plantes Méd. Phytothér.* **19**, 84–88.

Janot, M.-M., Guilhem, J., Contz, O., Venera, G. and Ciongo, E. (1979) Contribution à l'étude des alcaloides de la Valériane (*Valeriana officinalis* L.); Actinidine et naphtyridylméthylcétone, nouvel alcaloide. *Annales Pharmaceutiques Françaises* **37**, 413–420.

Johnson, R.D. and Waller, G.R. (1971) Isolation of actinidine from *Valeriana officinalis*. *Phytochemistry* **10**, 3335–3336.

Joshi, G.D., Viagya, A.S., Kulkarni, S.N. and Bhattacharyya, S.C. (1968) Terpenoids CXX: Components of Indian Valerian root oil Part II *Perf. Ess. Oil Rec.* **59**, 187–191.

Koch. U and Hölzl, J. (1985) The Compounds of *Valeriana alliariifolia;* Valepotrathydrines. *Planta Medica* 172–173.

Krepinsky, J., Romanuok, M., Herout, V. and Sorm, F. (1962) Structure of the Sesquiterpene Ketone Valeranone. *Collection of Czechoslovak Chemical Communications* **27**, 2638–2653.

Kucaba, W.E., Thies, P.W. and Finner, E. (1980) Isodidrovaltratum, ein Neues Valepotriat aus Valeriana vaginata. *Phytochemistry* **19**, 575–577.

Laufer, J.L., Seckel, B. and Zwaving, J. (1970) Composition of the Active Principles of Different Valeriana and Kentranthus species. *Pharmaceutisch Weekblad* **105**, 609–625.

Lin, L-J, Cordell, G.A. and Balandrin, M.F. (1991) Valerian-derived Sedative agents. I. On the Structure and Spectral assignment of the Constituents of Valmane Using the Selective INEPT Nuclear Magnetic Resonance Technique. *Pharmaceutical Research* **8**, 1094–1102.

Mannenstaetter, E., Gerlach, H. and Poethke, W. (1966) Phytochemical Stiudies on *Centranthus ruber. Pharmazie* **21**, 321–327.

Marekov, N.L., Popov, S.S. and Handjieva, N.V. (1983) Chemistry of Pharmaceutically-important Cyclomonoterpenes from some Valeriana plants in: *Chemistry and Biotechnology of Biologically-active Natural products*; 2nd International Conference, Budapest. Elsevier, Amsterdam, pp. 313–341.

Narayanan, C.S., Kulkarni, K.S., Vaidya, A.S., Kanthamani, S., Lakshmi Kumari, G., Bapat, B.V., Paknikar, S.K., Kulkarni, S.N., Kelkar, G.R. and Bhattacharyya, S.C. (1964) Terpenoids XLVI: components of Indian Valerian Root Oil. *Tetrahedron* **20**, 963–968.

Nikiforov, A., Remberg, B.and Jirovetz, L. (1994) Headspace Analysis of Valerian Roots (Radix Valerianae) *Scientia Pharmaceutica* **62**, 331–335.

Nishiya, K., Kimura, T., Takeya, K. and Itokawa, H. (1994) Sesquiterpenoids and iridoid glycosides from *Valeriana fauriei. Phytochemistry* **36**, 1547–1548

Nishiya, K., Tsujiyama, T., Kimura, T., Takeya, K., Itokawa, H., and Iitaka, Y. (1995) Sesquiterpenoids from *Valeriana fauriei. Phytochemistry* **39**, 713–714.

Oshima, Y., Hikino, Y; and Hikino, H. (1986) Structure of cyclokessyl acetate, a sesquiterpenoid of Valeriana fauriei. *Tet. Lett.* 27, 1829–1832.

Oshima, Y., Matsuoka, S. and Ohizumi, Y. (1995) Antidepressant Principles of Valeriana fauriei Roots. *Chem. Pharm. Bull.* **43**, 169–170.

Paknikar, S.K. and Kirtany, J.K. (1972) The stucture of *Valeriana wallichii* hydrocarbon. *Chem. Ind.* **21**, 803.

Pande, A. and Shukla, Y.N.(1993) A naphthoic acid derivative from *Valeriana wallichii. Phytochemistry* **32**, 1350–1351.

Pande, A., Uniyal, G.C. and Shukla, Y.N.(1994) Determination of chemical constituents of *Valeriana wallichii* by reverse phase HPLC. *Ind. J. Pharm. Sci.* **56**, 56–58.

Popov, S., Handzhieva, N.V., and Marekov, N. (1973) Halogen-containing Valepotriates isolated from Valeriana officinalis roots. *Doklady Bolgarskoi Akademii Nauk* **26**, 913–915.

Popov, S., Handzhieva, N.V., and Marekov, N.(1974) A new valepotriate 7-epideacetylisovaltrate from *Valeriana officinalis. Phytochemistry* **13**, 2815.

Rybal'chenko, A., Fursa, N and Litvinenko, V. Phenolic compounds of the Epigeal part of Valerian (1976) *Khim. Prir. Soedin.* **1**, 106–107 (**CA 85**, 59568z)

Rücker, G. (1979) Über die Wirkstoffe der Valerianaceen. *Pharmazie in Unserer Zeit* **8**, 78–86.

Rücker, G. and Tautges, J.(1976) β-Ionone und patchoulialkohol aus den unterirdischen teilen von Valeriana officinalis. Phytochemistry **15**, 824.

Rybal'chenko, A.S., Fursa, N.S. and Litvinenko, V.I.(1976) Phenolic compounds of the aboveground parts of Valeriana I Phenolcarboxlic acids and flavonoids. Khim. Proer. Soedin. 106–107 (**CA 85**, 59568).

Santos, M.S., Ferreira, F., Faro, C., Pires, E., Carvalho, A.P., Cunha, A.P. and Macedo, T. (1994) The amount of GABA Present in Aqueous Extracts of Valerian is Sufficient to Account for [^3H]GABA Release in Synaptosomes. *Planta Medica* **60**, 475–476.

Schneider, G. and Willens, M. (1979) Neue Abbauprodukte der Valepotriate aus Kentranthus ruber (L.) D.C. *Archiv der Pharmazie* **315**, 555–556.

Schulte, K.E., Rücker, G., and Glauch, G. (1967) Über einige Inhaltstoffe der Unterirdischen Teile von Nardostachys chinensis Batalin. *Planta Medica* **15**, 274–281.

Sietz, G. and Moddelmog, G. (1984) Synthese und Spektroskopische Eigenschaften von Stickstoffe- und Schwefelvarainten des Baldrinals. *Österr. Apoth. Ztg.* **38**, 50.

Stahl, E. and Schild, W (1969) Ein Charakteristisches Chromogenes Valepotriate ohne Dienstruktur in Valerianaceen. *Tetrahedron Lett.* 1053–1056.

Stahl, E. and Schild, W. (1971) Über die Verbreitung der Aequilibriend Wirkenden Valepotriate in der familie der Valerianaceen. *Phytochemistry* **10**, 147–153.

Stoll, A. and Seebeck, E. (1957) Die Isolierung von Hesperitinsäure, Behensäure and zwei unbekannten Säure aus Baldrian. *Liebigs Ann. Chem.* **603**, 158–168.

Stoll, A., Seebeck, E. and Stauffacher, D. (1957a) New Investigations on Valerian. *Schweizerische Apotheker-Zeitung* **95**, 115–120.

Stoll, A., Seebeck, E. and Stauffacher, D. (1957b) Isolierung und charakterisierung von bischer unbekannten inhaltstoffen aus dem neutraltiel des frischen baldrians. *Helv. Chim. Acta* **40**, 1205–1227.

Suzuki, H., Zhang, B.C., Harada, M., Iida, O. and Satake, M. (1993) Quantitative studies on terpenes of Japanese and European valerians. *Jap. Journal of Pharmacognosy* **47**, 305–310.

Thies, P.W. (1966) Über die Wirkstoffe des Baldrians 2, Zur Konstitution der Isovaleriansaureester Valepotriat, Acetoxyvalepotriatund Dihydrovalepotriat. *Tetrahedron Letters*, 1163–1170.

Thies, P.W. (1968a) Die Konstitution de Valepotriate. *Tetrahedron* **24**, 313–347.

Thies, P.W. (1968b) Linarin-isovalerianate, ein bischer unbekannte Flavonoid aus V. wallichii D.C.. *Planta Medica* **16**, 361–371.

Thies, P.W. (1970) Valerisodatum, ein Iridoidesteglycosid aus *Valeriana*-arten. *Tetrahedron Lett* 2471–2474.

Thies, P.W., Finner, E. and David, S. (1981) Konstitutiver Zuordnung der Acyloxysubstituentten in Valepotraten von C-13 Spektroscopie. *Planta Medica* **41**, 15–20.

Thies, P.W., Finner, E. and Rosskopf, F. (1973) Die Konfiguration des Valtratum und anderer Valepotriate. *Tetrahedron* **29**, 3213–3226.

Titz, W. (1983) Valepotriates and essential oil of morphologically and karyologically defined types of *V. officinalis* II Variation of some characteristic compounds of essential oil. *Sci Pharm* **51**, 63–67.

Torsell, K. and Wahlberg, K. (1967) Isolation, structure and synthesis of alkaloids from Valeriana officinalis L. *Acta Chemica Scandinavica* **21**, 53–62.

Tortarolo, M., Braun, R., Huebner, G.E. and Maurer, H.R. (1982) In vitro effects of epoxide-bearing valepotriates on mouse early haemopoietic progenitor cells and human T-lymphocytes. *Arch. Toxicol.* **51**, 37–42.

Trzhetsinskii, S.D., Fursa, N.S., Litvinenko, V.I., Postrigan, I.G.and Zaitsev, V.(1982) Phenolic compounds of the aboveground parts of Valeriana IV. Composition of flavonoids of three Central Asian species of Valeriana; Khim. Proer. Soedin. 255 (**CA 97**, 123929).

Ukida, T. (1944) Structure of kessoglycol. *Journal of the Pharmacological Society of Japan* **64**, 285–294.

Violon, C., Dekegel, D. and Vercruysse, A. (1984) Relation between valepotriate content and differentiation level in various tissues from Valerianeae. *J. Nat. Prod.* **47**, 934–940.

Vömel, A., Hölzl, J. and Fuckël, I. (1984) Ontogenese von *Valeriana officinalis* L. und *V. alliariifolia* Vahl. *Osterreichische Apother-Zeitung* **38**, 43–44.

Zaitsev, V.G., Fursa, N.S. and Zhukov, V.A. (1985) Flavonoids and valepotriates of valerian VIII Valeriana spryginii. Khim. Proer. Soedin. 568–569 (**CA 104**, 85447).

3. THE PHARMACOLOGY AND THERAPEUTICS OF *VALERIANA*

JOSEF HÖLZL

*Institüt für Pharmazeutische Biologie der Philipps Universität,
3550 Marburg, Germany*

CONTENTS

Antifungal Activities
Animal Attractant Properties

TOXICOLOGY OF VALERIAN

General Considerations
Cytotoxicology of Valepotriates

CONCLUSION

INTRODUCTION

Extracts of various species of *Valeriana* are used in the traditional medicine of many parts of the world where they are endemic. The major traditional use is for purposes which can be classified as tranquillising or sedative but they are also used as gastrointestinal sedatives, poison antidotes, deodorants and for treating urinary tract disorders (Hobbs, 1989). Species of *Valeriana* are official in many national pharmacopoeias and are also used in many proprietary phytotherapeutic preparations sold to promote sleep and reduce tension.

This chapter deals with the scientific evidence for the reputed activity of *Valeriana* and its constituents.

THE SEDATIVE EFFECT

The primary use of *Valeriana* species and their extracts in pharmacy and phytotherapy is as a sedative or tranquillizer and to help induce sleep and it is this purpose which is largely discussed below. The definition of the terms 'sedative' and 'tranquillizer' causes some confusion and they are often used interchangeably. However, as a general rule, a sedative causes a reduction in motor activity and mental disturbance often leading to sleep whilst a tranquillizer reduces mental disturbance without impeding motor function and mental alertness significantly.

Pharmacological studies were undertaken as long ago as 1907 which demonstrated that *V. officinalis* extracts possessed sedative effects experimentally (Chevalier,1907). Many subsequent studies have confirmed these early findings but the identity of the compounds responsible has been a matter of controversy which has still not been fully resolved.

Pharmacological Methods Of Detection

Many pharmacological test methods are available for investigating the sedative effect of substances. Methods used in the case of *V. officinalis* included the motility reduction of laboratory rodents, the lengthening of thiopental sleep, neurophysiological methods including measurement of the pharmaco-EEG, the desoxyglucose technique with measurement of the glucoseum sediment in different brain structures, and the procedure of receptor-binding studies for the tracing of effective substances. Total extracts of *V. officinalis* and also individual constituents of the plant have been investigated in these tests.

Measurement of motility

The measurement of motor activity of rats and mice is a classic experimental model for the investigation of the depressant action of substances on the CNS. There are various models for the measurement of motility. In a light barrier cage beams of light are interrupted by movements and these interruptions are registered. In a vibration cage movement produces vibrations which are registered. In the activity monitor lines of electric field are set up in the cage; movement causes these to be interrupted and this is measured as an impulse (Houghton, 1988).

Electroneurophysiological investigations

Electroencephalograph (EEG) output from the brain alters when CNS depressants are introduced and so the influence of doses of extracts or compounds on different parts of the brain can be studied by implantation of electrodes in the appropriate areas. The EEG output is recorded as a pattern of waves of electrical activity which are characteristic for different parts of the brain and which vary according to psychiatric disease states, alertness and consciousness.

The Sokoloff method

An extraordinarily elegant method for detecting central neuronal activity was published during the seventies, the desoxyglucose-technique of Sokoloff *et al.* (1977). The principle of this method is to determine quantitively the glucoseum sediment *in vivo* in different brain structures by using ^{14}C-labelled 2-desoxy-D-glucose. As the brain satisfies its energy requirement almost entirely through glucose and the individual brain structures only assimilate as much glucose as is necessary for their activity, a measure for the neuronal activity of the respective brain areas can be obtained through the quantitive determination of the local cerebral glucoseum sediment. These methods can also be used to localise and quantify the effects of medicinal substances on the brain *in vivo*. The desoxyglucose-technique has already proved its worth many times in the determination of the crucial effect of different medicinal substances.

Receptor-binding studies

In the majority of cases medicines have their effect because of their chemical structure. This presupposes that a specific structure is present in the organism that recognises the medical substance. This structure is often a receptor for endogenous hormones, transmitters or mediating substances for which the medicinal substance works as a substitute or antagonist.

Receptors are usually large proteins often localised on the outer surface of the cell (cell membrane). The chemical fit gives rise to connections and reactions, e.g. to a deformation, as a result of which an ion channel opens and thus produces an electric impulse. The contact of the agonist or antagonist frequently leads to the advent of a secondary messenger (e.g. cAMP, diacylglycerine, inositol phosphate, Ca^{2+}) and subsequent reaction.

Medicinal substances can be either agonists or antagonists; either they trigger a reaction or they produce a receptor blockade. About 40 receptor models are available for the search for the chemical working principle of a mixture of substances effecting the CNS.

Receptor-binding studies have been developed over the last ten years and are now extensively used in the investigation of biologically-active substances. The Radio Receptor Assay (RRA) method has indeed the disadvantage of low specificity as every substance with sufficient affinity to the receptor is driven to the radioligands. This fact, however, has the consequence that RRA can be employed for the establishment of a whole pharmacological class of effective substances.

Pharmacological Studies On Valeriana Extracts

Animal studies

Using motility experiments, Gstirmer and Kind (1951) indicated that the activity of the total extract could not fully be accounted for by the volatile oil content and other work confirmed these findings (Gstirmer and Kleinbauer, 1958). Stoll *et al.* (1957) were unable to explain the activity of a sample under investigation in terms of of the bornyl esters which were the major components of its volatile oil. However, the reality of the sedative effect of tinctures was demonstrated by reduction of the motility and an alteration of the reflex responses of mice (Kiesewetter and Muller, 1958).

Work in the United States to identify the active components in a fractionated alcoholic extract using prolongation of barbital-induced sleeping time and hypotensive effects was unsuccessful although active fractions were found (Rosecrans *et al.*, 1961).

Lecoq *et al.* (1963) showed that an extract of *V. officinalis* had an encephalic action, suppressed provoked metachronoses and , in sufficient amounts, neutralised the effects of alcohol.. Subsequently it was found that the influence of the extract over the effects of alcohol was more like that of chlorpromazine than hypnotic drugs (Lecoq *et al.*, 1964).

The volatile oil component of *Valeriana* spp. undoubtedly makes a major contribution to the sedative activity but there is much evidence that the non-volatile valepotriates and their metabolites also play a significant part since the isolation of the valepotriates (Thies, 1966) and the subsequent demonstration of their sedative activity (von Eickstedt, 1969) helped explain the discrepancy between the observed activity of the root tinctures and the smaller calculated effect based on the volatile oil components present (see below).

However, the total sedative activity of extracts of *Valeriana* species could not always be explained by the presence of valepotriates. Japanese samples of *V. officinalis* containing low amounts of valepotriates showed a greater affect on hexobarbital-induced sleeping time than Chinese and Nepalese samples containing high amounts of valepotriates (Hikino *et al.*, 1980). An explanation for this was provided by noting the high levels of volatile oil in the Japanese roots and the strong sedative effect of the kessyl derivatives comprising the bulk of the volatile oil.

Wagner *et al.* (1980) showed that, although the lipid-soluble fraction at 10mg dose gave a 50% reduction in the motility of mice, there was also some activity with the water-soluble fraction which gave up to 30% reduction in motility at 100mg dose. Another study in mice showed that intraperitoneal injection also depressed the CNS activity, and that oral administration had a greatly reduced effect (Veith *et al.*, 1986).

Motility experiment results obtained by the evaluation of a commercially available valerian root extract (Valdispert) revealed pronounced sedative properties in the mouse with respect to a reduction in motility and an increase in the thiopental sleeping-time (Leuschner *et al.*,1993). A direct comparison with diazepam and chlorpromazine revealed a moderate sedative activity for the tested extract. The extract showed only weak anti-convulsive properties.

A recent study on the ethanolic extract of the roots of *V. officinalis* compared its neuropharmacological effects with those of diazepam and haloperidol (Hiller and Zetler, 1996). Spontaneous motility, nociception and body temperature were not modified but anticonvulsant activity against a standard dose of picrotoxin was observed and barbitone-induced sleeping time was increased.

The psychotropic effects of roots of Japanese valerian, *V. faurei*, were compared with those of diazepam and imipramine (Sakamoto *et al.*,1992). Both the ethanolic extract of the root and diazepam significantly prolonged hexobarbital-induced sleeping times in mice. Spontaneous ambulation and rearing were significantly decreased by the *V. officinalis* extract, but kessyl glycol diacetate **1** and diazepam significantly increased ambulation. Diazepam significantly decreased approach-avoidance conflict in mice in a water-lick conflict test, but *V. officinalis* extract and KGD did not. By contrast, *V. officinalis* extract and imipramine significantly inhibited immobility induced by a forced swimming test in rats, but did not increase spontaneous motor activity during an open field test just before the forced swimming test. In addition, *V. officinalis* extract and imipramine significantly reversed reserpine-induced hypothermia in mice. These results indicate that *V. officinalis* extract acts on the central nervous system and may be an antidepressant.

A methanol extract of the roots of *Valeriana faurei* exhibited antidepressant activity in mice (Oshima, 1995). The extract was fractioned and alpha-kessyl alcohol **2** isolated as the active principle. The antidepressant activity of some guaiane and valerane types of sesquiterpenoids in the active fraction was also evaluated.

A study on the aqueous extract of *V. adscendens* showed a reduction in locomotor activity and coordination of movement when injected into mice (Capasso *et al.*, 1996). Prolongation of barbitone-induced sleep was also noted but there was no change in pain threshold levels and no effect on isolated guinea pig ileum. These results led the authors to conclude that the extract exhibited neuroleptic properties.

Electroneurophysiological investigations

An influence by *V. officinalis* extract and valtrate **3** was measured in the course of a neurophysiological investigation (Holm *et al.*, 1988). Stray cats were implanted with long-lasting electrodes. The dosage consisted of 5 or 20mg of valtrate **3** or isovaltrate **4** per kg and 100 or 250mg. per kg of *V. officinalis* extract. No regular changes of the contactial and subcortical EEG were apparent in these investigations. The muscular tone was reduced in **30–40%** of the cases. Among the electrically triggered reaction potentials, the hippocampial response to stimulation of the amygdaloid body was magnified with the local substances. The increase in amplitude of the amygdalo-hippocampial stimulation response had similarities with the effects of impramine and kawain among others. These effects can be ascribed clinically most nearly to the thymoleptic attributes. A sedation is unlikely, as is a vigilance reduction; the results, on the contrary, point to an improvement of the sensory and mental functions.

An objective demonstration of the efficacy of a plant sedative (250mg *V. officinalis* extract plus 60mg hop extract) was carried out in a clinical study by means of quantitive EEG measurement (Schellenberg, 1995). Before treatment the patients manifested measurable hyperreactions, quantifiable neurophysiologically as reduced occipital alpha-performance. It was difficult, if at all possible, for the brain to produce a relaxed waking state. The neurophysiological function deficiency was normalised after two weeks of medication. In a quantitive EEG the neurovegetative patients showed a change above all in the alpha-frequency range of between 8 and 12Hz. This result was also manifested in an improvement of the clinical condition, confirmed by a psychopathological rating scale.

The Sokoloff method

Grusla *et al.* (1986) carried out a series of experiments which used the Sokoloff method to investigate the effect of *V. officinalis* extract on glucose consumption in different areas of the brain. A marked change in the local cerebral glucoseum sediment in contrast to the control sediment was manifest in particular with 50mg *V. officinalis* extract per kg. Significant reductions in the glucose consumption were measurable above all in the different regions of the cortex, in the region of the limbic system and in the structure of the rhombencephalon. These results speak for an inhibiting effect on the neuronal activity and are reconcilable with a sedative effect of *V. officinalis*. Experiments with isolated single substances (valtrate **3**, didrovaltrate **4**, valerenic acid **6**, valeranone **7**) resulted, however, in no significant reduction of the glucoseum sediment; so the nature of the substances responsible for this activity is still not known (Kriegelstein *et al* 1986).

Receptor binding studies

γ-Aminobutyric acid (GABA) is an important neurotransmitter which mediates the inhibition of stress and anxiety. Levels of GABA, and hence its activity, are decreased by its uptake into presynaptic terminals. Aqueous extracts of *V. officinalis* have been shown to decrease GABA uptake in isolated synaptosomes by using radiolabelled GABA (Santos *et al.*, 1994a). 50% inhibition was given by 1μg/mL extract and 100% inhibition by 8μg/mL. GABA previously accumulated in the synaptosomes was also released independent of calcium ion concentration. A later report showed that the release of GABA from the synaptosomes stores was probably due to the concentration of GABA in the *V. officinalis* extract (Santos *et al.*, 1994b). The high levels of glutamine present could also be metabolised *in situ* to GABA and thus could contribute to the overall GABAergic effect.

The same group of workers reported more fully on the interaction of extracts of *V. officinalis* on GABA$_A$ receptors (Cavadas *et al.*, 1995). [^{3}H] muscimol binding techniques on rat brain cortices were used and it was found that both aqueous and alcoholic extracts displaced the muscimol. Both extracts had a similar level of GABA and, when a mixture of the amino acids was made of the same amounts found in the extracts, it displaced muscimol in a very similar way. Valerenic acid was not seen to have this effect . These results indicated that GABA present in the extracts was the major component binding to the receptors but its inability to cross the blood-brain barrier implies that any effect of the GABA observed *in vivo* would be peripheral rather than on the CNS.

Clinical Studies

Most scientific investigations of the efficacy of phytopharmaceuticals in the treatment of nervous conditions and sleeplessness have investigated *V.officinalis* extracts. All in all a subjective improvement in nervous conditions and in sleep quality can be established from the placebo-controlled double-blind studies and multi-centre studies.

The value of a preparation containing a mixture of extracts of *V. officinalis* and hops (*Humulus lupus* L.) for reducing stress in traffic was demonstrated (Moser, 1981). A double blind crossover trial was used and significant improvement was seen in both subjective experiences of stress and in objectively-measured reaction times.

Leathwood *et al.* (1982, 1985) have demonstrated a decrease in the time taken to fall asleep when an aqueous extract was used. There was also a subjective improvement in sleep quality and no "hangover" in the morning. An aqueous extract which contained only 0.01% w/v valepotriates and in which no volatile oil constituents were present indicated that some other substance may be present which improves sleep (Leathwood *et al.*, 1982, 1985). Its nature is still unknown although recent studies on the GABA content of *V. officinalis* suggest that this amino acid may exert some effect (Santos *et al.*, 1994a,b). In quantifying the effects of mild sedatives both physiological and subjective aspects of sleep must be taken into account. These were included in the study by Leathwood *et al.* (1982) who used a questionnaire analysis. The results showed that by subjective criteria *the V. officinalis* extract was sedative (i.e., it significantly decreased perceived sleep latencies and night awakenings, and improved sleep quality). The questionnaire also demonstrated that *V. officinalis* produced a significant decrease in subjectively-evaluated sleep latency scores and a significant improvement in sleep quality. The latter was most notable among people who considered themselves poor or irregular sleepers, smokers, and people who thought they normally had long sleep latencies. Night awakenings, dream recall and somnolence the next morning were relatively unaffected by valerian. In an EEG study on the same preparation the pattern of results tended to confirm the subjective evaluation (i.e. shorter mean sleep latency, increased mean latency to first awakening) but the changes did not reach statistical significance.

Another study was carried out with patients who complained of such symptoms as mental restlessness, delayed onset of sleep, frequent awakening and concentration difficulties. The medication was a pharmaceutical containing 45mg aqueous *V. officinalis* per dragee and 3 to 9 dragees were administered daily. The period of medication lasted for 10 or more days (Kamm-Kohl *et al.,* 1984). The success of the therapy was verified by observation with semi-quantitive scales throughout the test period. Towards the end of the test both patient and tester separately gave, on the basis of these observations, their assessment of the therapy results.

The effects of *V. officinalis* at two doses were investigated by computer analysis of sleep stages (sleep profiles) and psychometric methods (Gessner *et al.,* 1984). Both dosages showed a decrease of sleep stage 4 and a slight reduction of REM-sleep. On the other hand, a slight increase of sleep stage awake, 1 and 2 could be observed. A further increase of sleep stage 3 was identified. After application of 120 mg valerian, the frequency of REM-phases declined during the first half of the night, whereas during the second part of the night, a surplus appeared. Changes in the Beta-intensity of the EEG during REM-sleep showed a stronger hypnotic effect for the larger dose. Maximum effect was observed 2 - 3 hours after medication. Results of the mood scale were not different between the experimental conditions, which indicated no negative side-effects due to the drug or testing methods.

The effect of an aqueous extract of *V. officinalis* root on sleep was studied in two groups of healthy, young subjects (Balderer *et al.,* 1985). One group slept at home, the other in the sleep laboratory. Under home conditions, two different doses of *V. officinalis* extract reduced perceived sleep latency and wake time after sleep onset. Night-time motor activity was enhanced in the middle third of the night and reduced in the last third. The data suggested a dose-dependent effect. In the sleep laboratory, no significant differences from placebo were obtained. However, the direction of the changes in the subjective and objective measures of sleep latency and wake time after sleep onset, as well as in night-time motor activity, corresponded to that observed under home conditions. There was no evidence for a change in sleep stages and EEG spectra.

A further study (Schmidt-Voigt,1986) also showed a success rate of 35 to 66% for an improved ability to fall asleep. The *V. officinalis* preparation was similarly successful in improving sleep duration. In relation to these studies it should be noted that no better results were achieved using strong sedatives or psychotherapeutic drugs. It was remarkable that in these studies the patients reported an early effect of the drug, on average already after 1.6 to 1.8 days, particularly in cases of mental restlessness. Side effects such as stomach disorders, headaches and itching were noted but occurred only rarely.

A double blind test was carried out on a preparation containing primarily sesquiterpenes (Lindahl and Lindwall, 1989). When compared with placebo it showed a good and significant effect on poor sleep. 44% reported perfect sleep and 89% reported improved sleep from the preparation. No side effects were observed.

The effect of acute and repeated treatment with a *V. officinalis* extract on objective and subjective measures of sleep was studied (Schulz *et al.,* 1994). Polysomnography was conducted in 14 elderly poor sleepers on three nights, at one-week intervals. Subjects in the *V. officinalis* group showed an increase in slow-wave sleep (SWS) and a decrease in sleep stage 1. Density of K-complexes was increased under active treatment. There was no effect on sleep onset time or time awake after sleep onset, REM sleep was unaltered and there was also no effect on self-rated sleep quality. It was hypothesised that *V. officinalis* extract increases SWS in subjects with low baseline values.

Another multicentre study reported by Orth-Wagner *et al.* (1995) used a mixture of extracts from *V. officinalis*, Hops (fruits of *Humulus lupulus*) and Balm (*Melissa officinalis* herb). 225 patients who had trouble falling asleep because of nervous unrest were given the mixture for two weeks. In over 80% of the subjects significant subjective and objective improvements in ability to fall asleep, sleeping time and decrease in anxiety were noted.

Pharmacological studies on individual constituents

Valeranone sesquiterpenes

Nardostachys jatamansii DC. roots contain up to 3% v/w volatile oil and valeranone **7** is the major component present in the volatile oil. Arora and Arora (1963) showed that a hypotensive effect in mice was observed at a dose of 5 mg/kg valeranone **7** and a larger dose (100 mg/kg) significantly increased pentobarbitone-induced sleeping time when given both orally and by intraperitoneal injection to mice. It was shown that levels of 5-hydroxytryptamine and noradrenaline in the brain of the rabbit were reduced after administration of valeranone **7**. Such a reduction is a feature of some tranquillising drugs. The LD_{50} of valeranone **7** was found to be 580 ± 6 mg/kg thus indicating the relatively low toxicity of active doses.

Rücker *et al.* (1978) produced an even higher oral LD_{50} of greater than 3 g/kg for valeranone **7** but the hypotensive effect in rats was found to be fairly weak. In addition a prolongation of sleep induced by barbiturate was observed. A tranquillising effect was seen when the electric shock avoidance test was used but valeranone **7** at a dose of 31.6 mg/kg was not so potent as a 10 mg/kg dose of chlorpromazine.

Work carried out by Hendriks *et al.* (1981) showed that valeranone **7** at a dose of 100mg/kg gave a range of responses typical of sedative action. In a later study an antispasmodic effect was observed both *in vivo* and *in vitro* on guinea pig gut with doses of 20 mg/kg. Results of test carried out indicated that the action of valeranone **7** was musculotropic rather than an interaction with the autonomic nervous system (Hazelhoff *et al.*, 1982).

Kessane sesquiterpenes

The kessane derivatives had been known for some time as chemical curiosities but it was not until 1973 that tests showed that kessoglycol diacetate **1**, a major component of the oil of *V. officinalis* var. *latifolia*, decreased the motility and prolonged the hexobarbitone-induced sleeping time in rats (Takamura *et al.*, 1973). The LD_{50} was high being greater than 5 g/kg in mice. Subsequent structure-activity studies using several kessyl sesquiterpenes indicated that the kessyl alcohol 8-acetate **8** was more potent at prolonging sleep than the 2,8-diacetate **1** but had little effect on the motility of mice (Takamura *et al.*, 1975a). Butyl analogues prepared synthetically were also found to be more potent than the acetates (Takamura *et al.*, 1975b).

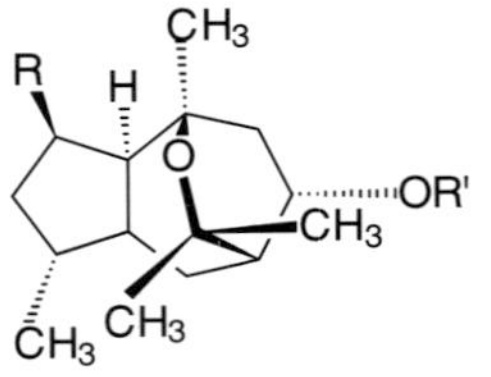

1 R, R' = OOCCH$_3$ Kessyl glycol diacetate
2 R, R' = H Kessyl alcohol
8 R = H, R' = OOCCH$_3$ Kessyl 8-acetate

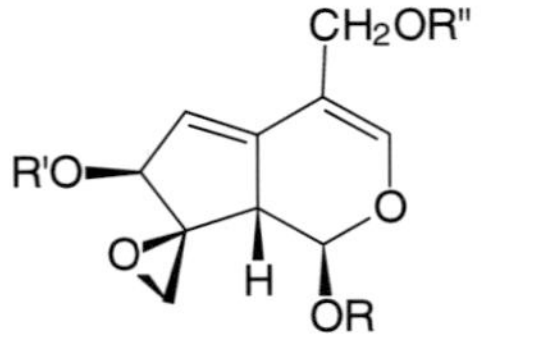

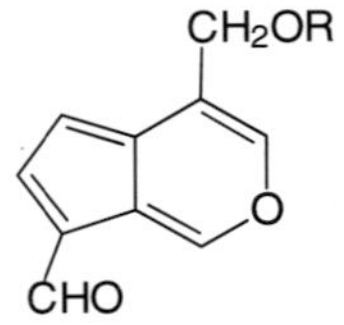

	R	R'	R"	
3	Iv	Iv	Ac	Valtrate
10	Aiv	Iv	Ac	Acevaltrate
4	Iv	Ac	Iv	Isovaltrate

	R	
11	Ac	Baldrinal
12	Iv	Homobaldrinal

Ac =CH$_3$CO—

Iv =

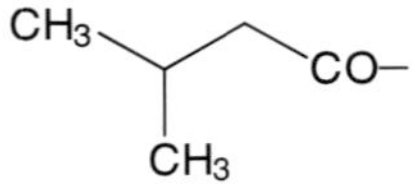

Aiv =

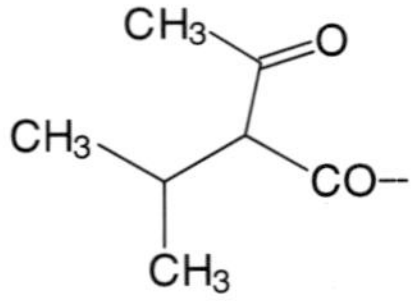

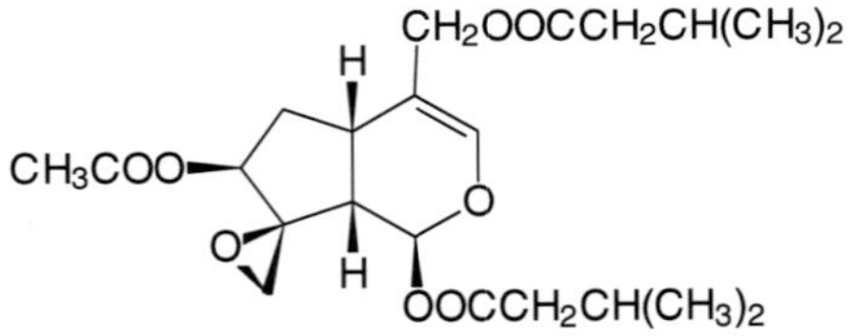

5 Didrovaltrate

Figure 1

Valerenic acid and related sesquiterpenes

Valerenic acid **6** was first isolated by Stoll *et al.* (1957) and shown to have a sedative effect on frogs and a spasmolytic effect on guinea pig gut. No further studies were performed until Hendriks *et al.* (1981) investigated several components of the volatile oil of *V. officinalis* for various responses indicating sedative activity on groups of mice. Doses of 50 mg/kg for valerenic acid **6** and valerenal **9** showed significant effects including decreased motor activity, decreased rotarod performance and ataxia.

This work led to a more detailed investigation into the activity of valerenic acid **6** (Hendriks *et al.*, 1985). Its influence on the rotarod and traction performance of mice was compared with chlorpromazine, diazepam and pentobarbital as well as its on pentobarbital-induced sleeping time and on spontaneous motor activity of mice. Significant reduction of locomotor activity was shown by doses of 60 mg/kg which adversely affected performance on the rotarod and in the traction experiment. The profiles for these tests indicated that valerenic acid **6** resembled pentobarbital and is likely to have a general central depressant activity rather than a muscle relaxant or neuroleptic effect. Large doses of valerenic acid (>400 mg/kg) caused strong peripheral bleeding, convulsions and death. Hiller and Zetler (1996) demonstrated that valerenic acid **6** reduced convulsions induced by picrotoxin.

Riedel *et al.* (1982) showed that valerenic acid **6** inhibited the enzyme system catalysing breakdown of GABA in the brain. The net effect is thus an enhanced level of GABA which is associated with sedation and a decrease in CNS activity.

Valerenic acid is therefore thought to be an important CNS-depressant constituent of the oil of those samples of *V. officinalis* which contain large amounts of this type of sesquiterpene.

Valepotriates

The isolation of the valepotriates (Thies, 1966) led to much interest into their activity and this has led to the widespread use in Europe of standardised mixture of valepotriates as a mild sedative. The pharmacology of these compounds was first reported by von Eickstedt and Rahman (1969) although some clinical testing had previously been carried out (Stocher, 1967).

Several experiments have been carried out using the commercial mixture containing 15% valtrate **3**, 80% didrovaltrate **5** and 5% acevaltrate **10**. Tests on mice using the running wheel demonstrated that the valepotriates had a tranquillising effect at doses of 31 mg/kg given orally but this effect was less than that given by 10 mg/kg chlorpromazine. On the other hand the rotarod test showed an improvement in co-ordination with the animals given valepotriates compared to those given chlorpromazine. Behavioural tests on cats showed no decrease in reactivity but decreases in restlessness, anxiety and aggressiveness.

An interesting experiment was carried out which compared the interaction of valepotriates, diazepam and chlordiazepoxide with alcohol (von Eickstedt, 1969). The addition of valepotriates to alcohol given to mice lessened the impairment of rotarod performance and somewhat prolonged anaesthesia. The toxicity of the alcohol was increased by addition of diazepam but not valepotriates.

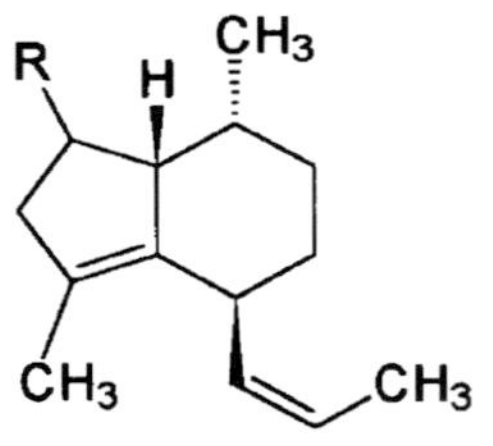

6 R = COOH Valerenic acid   7 Valeranone
9 R = CHO   Valerenal

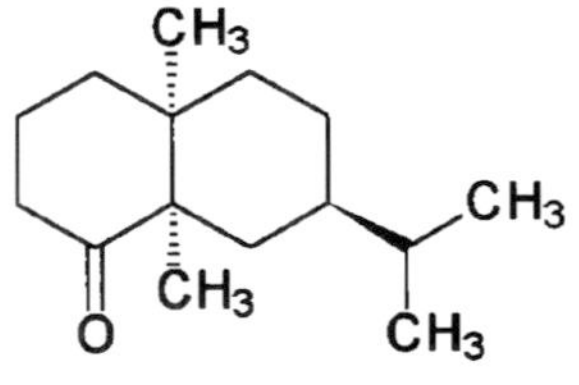

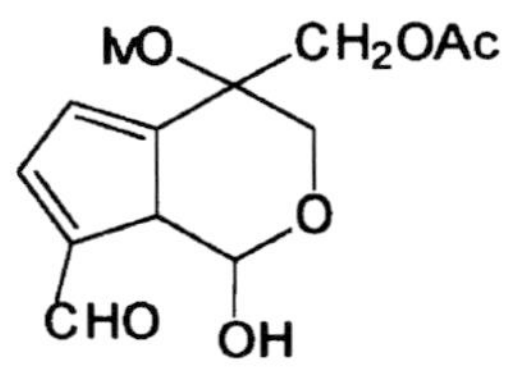

13 Valtroxal 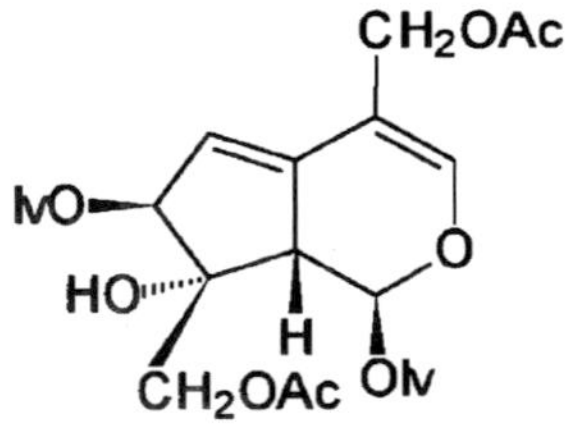   14   Acetoxyvalthydrine

Figure 2

Tests on the spontaneous motility of mice showed little extra reduction when the alcoholic extract of *V. officinalis* (containing valepotriates) was fortified with more volatile oil (Wagner *et al.*, 1980). This indicated that the valepotriates play the major part in sedative action. However, the composition of the oil of *V.officinalis* is notoriously variable and, consequently, so is its activity and it would be unwise to draw too many conclusions from this work. The same authors compared the effects of the valepotriates and their decomposition products such as baldrinal **11**.

Homobaldrinal **12** was found to have a greater effect on spontaneous motility than valtrate **3** and isovaltrate **4** at doses of 100 mg/kg given orally. It was therefore postulated that the hydrolysis products were the most active forms. These compounds seem to be metabolised quite rapidly in the gut as experiments with didrovaltrate **5** show (Wagner and Jurcic, 1980). Other decomposition products of valtrate **3** gave a sedative effect when injected into mice intraperitoneally (Schneider and Willems, 1982). Recent work showed that the most potent reduction of spontaneous motility was given by valtroxal **13**, one of the degradation products of didrovaltrate, rather than didrovaltrate **5** itself (Veith *et al.*, 1986).

These tests closely identified the sedative activity of Valerianaceous plant extracts with the valepotriates. However it is important to note that a decrease in activity of the tincture after a short storage time had been noted in very early work (Macht and Ting, 1921) and corresponds with low levels of valepotriates in traditional preparations of *V. officinalis* which have been left in storage for some time (Bounthanh *et al.*, 1980). The valepotriates hydrolyse quite quickly and cannot be detected after 60 days (Adzet *et al.*, 1975). The efficiency of extracts of *Valeriana* which have been stored for long periods of time is therefore questionable if valepotriates are the major active components present.

Recent concern about the alkylating potential of the epoxide group has prompted the investigation of valepotriates that do not contain it. Thus acetoxyvalthydrine **14**, which lacks the epoxide, was compared with valtrate **3** for its effect on the activity of mice (Hölzl and Fink, 1984). Valtrate **3** gave a significant effect in oral dose of 0.5 mg/kg but acetoxyvalthydrine was less active, needing a dose of 4mg/kg.

There is still uncertainty as to how the valepotriates exert their effect centrally. Electrophysiological studies on cats using electrodes to monitor activity in different areas of the cortex and sub-cortex of the brain indicated that the valepotriates act on the amygdaloid body (Hölm, 1984). Valtrate **3** and isovaltrate **4** showed similarity to some antidepressants in being thymoleptic whilst didrovaltrate **5** seemed to inhibit efferent impulses to the hippocampus in a similar way to the benzodiazepines, thus showing tranquillising properties.

Other recent work using perfused rat brain has shown that changes in the EEG pattern were produced in a dose-related fashion by valtrate **3** with a reduction in beta activity and an increase in theta and delta frequencies (Fink *et al.*, 1984).

Lignans

A newly detected substance, hydroxypinoresinol **15**, a lignan, showed the highest binding at the 5-HT receptor and an insignificant binding at the benzodiazepam receptor when *Valeriana* compounds were subjected to receptor-binding studies (Bodesheim *et al*, 1995).

STUDIES FOR OTHER ACTIVITIES

Spasmolytic Effects

The use of extracts of *Valeriana* in traditional medicine for gastrointestinal hypermotility and associated conditions such as diarrhoea has been validated to some extent by *in vitro* pharmacological investigations.

The valepotriates have been shown to exert a spasmolytic effect, first demonstrated for valtrate **3** and didrovaltrate **5** by Wagner and Jurcic (1979). The commercial mixture of valepotriates had an effect stronger than the same dose of papaverine but the individual compounds were less strong even at higher doses. Further investigations by Hazelhoff *et al.* (1982) revealed that valtrate **3**, isovaltrate **4** and didrovaltrate **5** probably act as

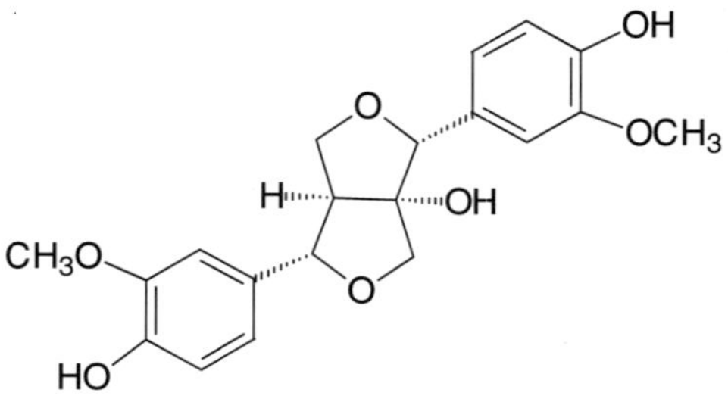

15 Hydroxypinoresinol

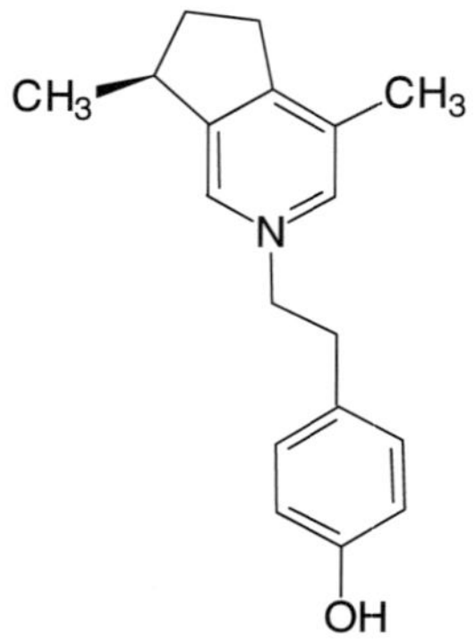

16

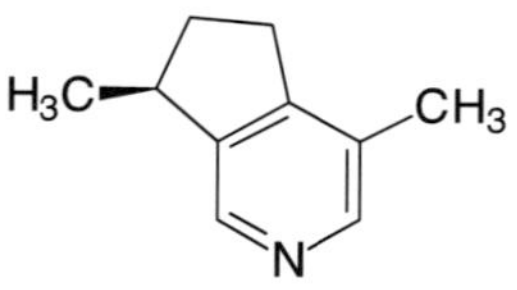

17 Actinidine

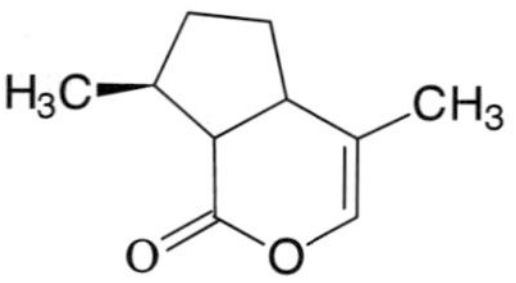

18 Nepetalactone

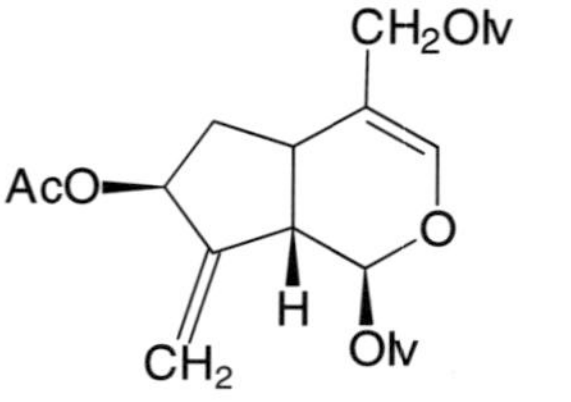

19 Deoxidodidrovaltrate

Figure 3

musculotropic agents. Their action may be due to an influence on the entry of Ca^{2+} ions or on their binding to the muscle.

During the course of receptor-binding studies several substances from *V. officinalis* were isolated that bind specifically to receptors. Nearly all the separate compounds from the group of valepotriates were constructed and tested on different receptors. The greatest activity among this group of substances was to be found at the dopamine receptor where valtrate **3** had a binding constant of IC 1.8×10^{-6} whilst the isomeric compound; isovaltrate was bound considerably less (Godau 1991).

Effect Of Valepotriates On The Complement System

The complement system comprises part of the immune system in the body and has been the subject of investigation for many plant extracts used in traditional medicine. When the valepotriates were tested all but one showed an inhibitory effect on the alternative synthesis route in the complement system of the serum (van Meer, 1984). Didrovaltrate **5** had the greatest activity whereas the decomposition products like baldrinal **11** were less active. This activity indicated a possible use for these compounds in some auto-immune diseases.

Cholinergic Effects

Little is known about the activity of the alkaloids of *V. officinalis* which are only minor constituents. One of the alkaloids **16** isolated by Torsell and Wahlberg (1967) showed a high degree of cholinesterase activity. This is a derivative of actinidine **17**, also found from *Valeriana* spp., which has a cholinergic effect.

Antifungal Activity

Ten valepotriates have been shown to exhibit activity against the plant pathogen fungus *Cladosporium cucumerinum* in a bioautographic assay on thin layer chromatography. (Fuzzati *et al.*, 1996). Valtrate **3** was active at 1μg and also showed significant inhibition at low concentrations on other fungi such as *Aspergillus fumigatus*, *Candida albicans* and *Trichophyton mentagrophytes*. Valtrate **3** and didrovaltrate **5** were also found to be active against other plant pathogens and the valepotriates show promise as a new group of fungicidal compounds.

Animal Attractant Properties Of *Valeriana*

Some species of *Valeriana* are noted for their property of attracting animals, particularly cats, although rats and animals related to dogs have also been mentioned in this respect. The alkaloid actinidine **17** has been postulated as the attractant since it also occurs in *Actinidia* , another cat-attractant genus (Tucker and Tucker, 1988). Nepetalactone **18** is found as a minor constituent of the volatile oil of some Valeriana species and this compound is considered to be the major agent responsible for the characteristic effect of catnip, *Nepeta cataria* (Bicchi *et al.*, 1983). It is also worth noting that isovaleric acid, responsible for the characteristic odour of stored *Valeriana* spp., is a component of the anal gland secretion of some members of the cat family and is associated with mating behaviour (Lissàk, 1962).

TOXICOLOGY OF VALERIAN

General Considerations

Numerous studies have idicated that aqueous and alcoholic extracts of *V. officinalis* have a high LD_{50} value. A recent case of overdose where the patient had ingested almost 25g of powdered *V. officinalis* root in capsule form demonstrated only mild symptoms (Willey *et al.*, 1995). These symptoms included fatigue, abdominal cramps and tremor but all of them disappeared within 24 hours. Four cases of hepatitis incurred by regular users of herbal sedatives consisting of a mixture of herbs including *V. officinalis* has been widely cited (Macgregor *et al.*, 1989). The evidence for the *V. officinalis* being the causative agent is circumstantial and may be due to the other components present rather than the *V. officinalis* (Perharic *et al.*, 1994).

Cytotoxicity Of The Valepotriates

Comparison of the structure of the valepotriates with the cytotoxic substance 6,20-epoxylathyrol-3, 5-dibutyrate revealed that the two types of molecule shared similar features (Bounthanh *et al.*, 1981). As a consequence valtrate **3**, didrovaltrate **5** and baldrinal **11** were tested *in vitro* on isolated liver cell tumours of rats for their cytotoxicity. Valtrate **3** was twice as toxic as didrovaltrate **5** and eight times as toxic as baldrinal **11**. 24 hours exposure of cells to 33mg valtrate per ml led to their total mortification. In further studies valtrate **3**, didrovaltrate **5** and deoxido-didrovaltrate **19** were tested on the same cell line for their cytotoxic activity. While 20 mM valtrate **3** led to cell death 48h after incubation, 40% of the cell population survived the dose of didrovaltrate **5** and deoxido-didrovaltrate **19**. Valtrate **3** was twice as toxic as didrovaltrate **5** .

The epoxide group present in most of the valepotriates is a common feature of alkylating agents which are thereby cytotoxic. Braun *et al.* (1982) demonstrated this alkylating ability and showed that valtrate **3** and didrovaltrate **5** alkylated 4-(*p*- nitrobenzyl)-pyridine at the same rate as the active substances epichlorhydrin and *N,N*-dimethyl *N*-(2-chloroethyl)amine. A time- and dose- related inhibition of incorporation of ^{14}C-thymidine in the DNA of Ehrlich ascites carcinoma cells shown by the two valepotriates implied a process of covalent binding

Bounthanh *et al.*, (1983) studied the interaction between these compounds and DNA synthesis in cultured hepatoma cells. Valtrate **3** had a rapid and extensive inhibitory effect on the incorporation of [3]H- thymidine and ^{3}H-leucine at concentrations as low as 20 mM and these amounts caused death of the cells but didrovaltrate **5** and deoxidodidrovaltrate **19** were less active but nevertheless showed considerable cytotoxicity. Structure-activity considerations indicated that the C5-C6 double bond is important for cytotoxic activity and that the epoxide group is not essential.

Tortarolo *et al.* (1982) found the valepotriates to be cytotoxic against mouse bone marrow early progenitor cells *in vitro* and the *in vivo* effect of valtrate 3 on the same type of cell in mice has subsequently been tested (Braun *et al.,* 1984). In contrast to the *in vitro* results no significant effect could be observed when the drug was given orally or intraperitoneally even in doses of 1350 mg/kg or 65 mg/kg, respectively. The same report showed that 50 mg/kg valtrate 3 reduced the ability of the liver to metabolise ^{14}C- methacetin when given intraperitoneally but no significant reduction was observed with oral administration, even at doses as high as 1500 mg/kg.

These results indicate that *in vivo* toxicity might not be so great as *in vitro* tests imply because of poor absorption and/or distribution and of the drug.

The cytotoxicity of the valepotriates has aroused concern about the safety of preparations from *Valeriana*, particularly if long-term use is involved. Recent trends in research into the volatile oil components and investigation of species such as *V. alliariifolia* Troitzky, which contains non-epoxide valepotriates (Koch and Hölzl, 1985), indicate interest in minimising the potential toxic effects whilst seeking to preserve the use of Valerianaceous plant as acceptable sedatives.

CONCLUSION

Although considerable progress has been made in elucidating the chemical basis for the observed sedative activity of *Valeriana* spp., the situation is still far from being completely resolved. *Valeriana* species present an interesting facet of phytotherapeuticals *viz.* the total effect being due to a mixture of chemical types with various pharmacological activities. The constituents seem to act both centrally on the brain and peripherally in causing relaxation of smooth muscle. The central action is due, in part, to activity at the GABA receptors, possibly because of significant amounts of that substance in the plant extract.

The clinical evidence indicates that *Valeriana* is a relatively safe substitute for the benzodiazepines as a mild tranquillizer and also that it has potential in aiding onset of sleep. The large number of commercial preparations available throughout the world which contain *Valeriana* spp. or their extracts bear some witness to the reputation and efficacy of this genus.

REFERENCES

Adzet, T., Inglesias, J., San Martin, R. and Torrent, M.T. (1975) Étude de certains esters de *Centranthus ruber* et action pharmacodynamique de quelques-unes de ses preparations galeniques. *Planta Medica* 27, 194–198.

Arora, R.B. and Arora, C.K. (1963) Hypotensive and tranquillising activity of jatamansone (valeranone) a sesquiterpene from *Nardostachys jatamansi* DC. In: K.K. Chen and B. Mukerji (Eds.), *Pharmacology of Oriental Plants*. Pergamon, Oxford, pp. 51 – 60.

Balderer, G. and Borbely, A. (1985) Effect of valerian on human sleep. *Psychopharmacology* 87, 406–407.

Bicchi, C., Sandra, P., Schelfaut, M. and Verzele, M. (1983) Studies on the essential oil of *Valeriana celtica* L. *J. High Resolution Chromatogr. & Chromatogr. Commun.* 6, 213–215.

Bodesheim, U and Hölzl, J. (1995) Isolierung und Strukturaufklärung eines neuen Lignans aus Valeriana officinalis und Testung am Rezeptormodell. *Pharmazie in Unserer Zeit* **24**, 329–342.

Bounthanh, D., Misslin, R. and Anton, R. (1980) Activitée comparé de preparations galenique de Valeriane, *Valeriana officinalis*, sur les comportement de la souris. *Planta Medica* **39**, 241–242.

Bounthanh, C., Bergmann, C., Beck, J.P., Haag-Berrurier, M. and Anton, R. (1981) Valepotriates, a new class of cytotoxic and antitumor agents. *Planta Medica* **41**, 21–28.

Bounthanh, C., Richert, L., Beck, J.P., Haag-Berrurier, M. and Anton, R. (1983) The action of valepotriates on the synthesis of DNA and proteins of cultured hepatoma cells. *Planta Medica* **49**, 138–142.

Braun, R., Dittmar, W., Machut, M. and Weickmann, S. (1982) Valepotriate mit Epoxidstruktur - beatliche Alkylantein. *Deutsche Apotheker-Zeitung* **122**, 1109–1113.

Capasso, A., De Feo, V., De Simone, F. and Sorentino, L. (1996) Pharmacological effects of aqueous extract from *Valeriana adscendens*. *Phytotherapy Res.* **10**, 309–312.

Cavadas, C., Araújo, I., Cotrim, M.D., Amaral, T., Cunha, A.P., Macedo, T. and Fontes Ribiero, C. (1995) *In vitro* study on the interaction of *Valeriana officinalis* L. extracts and their amino acids on GABA$_A$ receptor in rat brain. *Arzneim. Forsch.* **45 (II)**, 753–755.

Chevalier, J. (1907) Pharmacodynamic action of a new alkaloid contained in the roots of fresh valerian. *Compte Rendu de la Société de Biologie* **144** 154–157.

von Eickstedt, K.-W. (1969) Die Beeinflussung der Alkohol-Wirkung durch Valepotriate. *Arzneimittel-Forschung* **19**, 995–997.

von Eickstedt, K.-W. and Rahman, S. (1969) Psychopharmakologische Wirkungen von Valepotriaten. *Arzneimittel-Forschung* **19**, 316–319.

Fink. C., Hölzl, J., Rieger, H. and Krieglstein, J. (1984) Wirkungen von Valtrat auf das EEG des isoliert perfundierten Rattenhirns. *Arzneimittel-Forschung* **34**, 170–174.

Fuzzati, N., Wolfender, J.L., Hostettmann, K., Msonthi, J.D., Mavi, S. and Molleyres, L.P. (1996) Isolation of antifungal valepotriates from *Valeriana capense* and the search for valepotriates in crude Valerianaceae extracts. *Phytochemical Analysis* **7**, 76–85.

Gessner, B., Klasser, M. and Volp, A. (1983) Long term effect of a valerian extract on sleep in persons with sleep disorders. *Therapiewoche* **33**, 55476–55558.

Godau, P. (1991) Analytik von Inhaltsstoffen aus Valeriana officinalis und deren pharmakologischen Testung mit RBS. *Dissertation*, Univ. of Marburg.

Grusla, D., Hölzl, J and Krieglstein, J. (1986) Valerian effects on rat brain. *Deutsch. Apoth. Ztg.* **126**, 2249–2253.

Gstirmer, F. and Kind, H.H. (1951) Chemical and physiological examination of Valerian preparations. *Pharmazie* **6**, 57–63.

Gstirmer, F. and Kleinbauer, E. (1958) Zur pharmakologische Prüfung der Baldrianwurzel. *Pharmazie* **13**, 415–420.

Hazelhoff, B., Malingré, T.M. and Meijer, D.K.F. (1982) Antispasmodic effects of valerian compounds: an *in-vivo* and *in-vitro* study on the guinea-pig ileum. *Archives Internationales des Pharmacodynamie* **257**, 274–287.

Hendriks, H., Bos, R., Allersma, D.P., Malingré, T.M. and Koster, A.S. (1981) Pharmacological screening of valerenal and some other components of the essential oil of *Valeriana officinalis*. *Planta Medica*, **42**, 62–68.

Henriks, H., Bos, R., Woerdenbag, H.J. and Koster, A.S. (1985) Central nervous system depressant activity of valerenic acid in the mouse. *Planta Medica* 28 – 31.

Hiller, K.O. and Zetler, G. (1996) Neuropharmacological studies on ethanol extracts of *Valeriana officinalis* L.: Behavioural and anticonvulsant properties. *Phytotherapy Res.* **10**, 145–151.

Hikino, H., Hikino, Y., Kobinata, H., Aizawa, A., Konno, C. and Ohizumu, Y. (1980) Study on the efficacy of oriental drugs 18: Sedative properties of *Valeriana* roots. *Shoyakugaku Zasshi* **34**, 19–24.

Hobbs, C. (1989) Valerian. *Herbalgram* **21,** 19–34.

Hölm, E. (1984) Wirkungen von Valtratum/Isovaltratum und Didrovaltratum auf der Hirnrinde und subkortikale Hirngebiete. Elektrophysiologische Untersuchengen an Katzen. *Osterreichische Apotheker–Zeitung* **38**, 45–46.

Hölzl, J. and Fink, C. (1984) Untersuchengen zur Wirkung der Valepotriate auf die Spontanmotilitat von Mausen. *Arzneimittel-Forschung* **24**, 44–47.

Holm, H., Kowallik, H., Reinecke, A., von Henning, G.E., Behne, F. and Scherer, H.D. (1980) *Med. Welt.* **31**, 982.

Houghton, P.J. (1988) The biological activity of valerian and related plants. *J. Ethnopharmacology* **22**, 121-142.

Kamm-Kohl, A.V., Jansen, W. and Brockmann, P. (1984) Modern Valerian therapy against nervous disorders in senium. *Med.Welt* **35**, 1450–1454.

Kiesewetter, R. and Müller, M. (1958) Zur Frage der 'sedative' Wirkung von Radix Valerianae. *Pharmazie* **13**, 777–781.

Koch, U. and Hölzl, J. (1985) The compounds of *Valeriana alliariifolia*; Valepotrathydrines. *Planta Medica* 172–173.

Krieglstein, J. and Grusla, D. Centrally-acting inhibitory components in Valerian. (1988) *Deutsch. Apoth. Ztg.* **128**, 2041–2046.

Leathwood, P.D., Chauffard, F., Heck, E. and Munoz-Box, R. (1982) Aqueous extract of Valerian root (*Valeriana officinalis*) improves sleep quality in man. *Pharmacology, Biochemistry and Behavior* **17**, 65–72.

Leathwood, P.D. and Chauffard, F. (1985) Aqueous extract of valerian reduces latency to fall asleep in man. *Planta Medica* 144–148.

Lecoq, R., Chauchard, P. and Mazoue, H. (1963) Chronaximetric experimental study of the action of some mineral or vegetable sedatives and the associated effects of ethyl alcohol, with or without disulfiram. *Comptes Rendu de l'Academie des Sciences*, Paris **257**, 1403–1406.

Lecoq, R. Chauchard, P. and Mazoue, H. (1964) Chronaximetric investivation of certain pschyotropic agents and their modification of the neural effects of theyl alcohol. 1. Sedatives, analgesic and hypnotics. *Therapie* **19**, 967–974.

Leuschner, J., Müller, J. and Rudmann, M. (1993) Characterisation of the central nervous depressant activity of a commercially available valerian root extract. *Arzneimittel. Forsch. Drug Res.* **43** 638–641.

Lindahl, O. and Lindwall, L. (1989) Double blind study of a Valerian preparation. *Pharmacol. Biochem. and Behaviour* **32**, 1065–1066.

Lissák, K. (1962) Olfactory-induced sexual behaviour in female cats. in XXII International Congress of Physiological Sciences, Leiden, Netherlands. *Lectures and Symposia II. Symposia XI-XII* 653–656.

Macht, D. and Giu Ching Ting (1921) Experimental inquiry into the sedative properties of some aromatic drugs and fumes. *Journal of Pharmacology and Experimental Therapeutics* **18**, 261–372.

van Meer, J.H. (1984) Plantaardige stoffen met een effect op het complementsysteem. *Pharmaceutisch Weekblad* **119**, 836–942.

Moser, L. (1981) Arzneimittel bei Stress am Steuer? *Deutsche Apotheker-Zeitung* **121**, 2651–2654.

Orth-Wagner, S., Ressin, W.J. and Friederich, I. (1995) Phytosedative for sleeping disorders containing extracts from valerian root, hop grains and balm leaves. *Zeitschrift fur Phytotherapie* **16**, 147–152.

Oshima, Y., Matsuoka, S. and Ohizumi, Y. (1995) Antidepressant principles of *Valeriana fauriei* roots. *Chem. Pharm. Bull.* **43**,169–170.

Perharic, L., Shaw, D., Colbridge, M., House, I., Leon, C. and Murray, V. (1994) Toxicological problems resulting from exposure to traditional remedies and food supplements. *Drug Safety* **11**, 284–294.

Riedel, E., Hansel, R. and Ehrke, G. (1982) Hemmung des g-Aminobuttersaureabbaus durch Valerensaurederivate. *Planta Medica* **46**, 219–220.

Rosecrans, J.A., Defoo, J.J. and Youngken, H.W. (1961) Pharmacological investigation of certain *Valeriana officinalis* L. extracts. *Journal of Pharmaceutical Sciences* **50**, 240–244.

Rucker, G., Tautges, Z.J., Sienck, A., Wenzl, H. and Graf, E. (1978) Untersuchungen zur Isolierung und pharmakodynamischen Aktivitat ds sesquiterpens Valeranon aus *Nardostachys jatamansi* D.C. *Arzneimittel-Forschung* **28**, 7–13.

Sakamoto, T., Mitani, Y.and Nakajima, K. (1992) Psychotropic effects of Japanese valerian root extract. *Chem. Pharm. Bull.* **40**, 758–761.

Santos, M.S., Ferreira, F., Cunha, A.P., Carvalho, A.P. and Macedo, T. (1994a) An aqueous extract of valerian influences the transport of GABA in synaptosomes. *Planta Med.* **60**, 278–279.

Santos, M.S., Ferreira, F., Faro, C., Pires, E., Carvalho, A.P., Cunha, A.P. and Macedo, T. (1994b) The amount of GABA present in aqueous extracts of valerian is sufficient to account for [^{3}H]GABA release in synaptosomes. *Planta Med.* **60**, 2475–2476.

Schellenberg, R. (1995) Objecktiver Wirkungsnachweis eines pflanzlichen Sedativums mittels quantitativem EEG. *Apoth. J..* **17**, 55.

SchmidtVoigt, J. (1986) Treatment of nervous sleep doisorders and unrest with a sedative of purely vegetable origin. *Therapiewoche* **36**, 663–667.

Schulz, H., Stolz, C. and Müller, J. (1994) The effect of valerian extract on sleep polygraphy in poor sleepers; a pilot study. *Pharmacopsychiat.* **27**, 147–151.

Sokoloff, L., Reivich, M., Kenneda, C., des Rosiers, M.H., Patlak, C.S., Pettigrews, K.D., Sakurada, O. and Shinohara, M. (1977) The [14C] deoxyglucose method for the measurement of local cerebral glucose utilisation: theory, procedure and normal values in conscious and anaesthetised albino rats. *J. Neurochem.* **28**, 897–907.

Schneider, G. and Willems, M. (1982) Weiterer Erkenntnisse uber die Abbauprodukte der Valepotraite aus *Kentranthus ruber* (L.) DC. *Archiv der Pharmazie* **315**, 691–697.

Stoll, A., Seebeck, E. and Stauffacher, D. (1957) New investigations on Valerian. *Schweizerische Apotheker-Zeitung* **95**, 115–120.

Takamura, K., Kakimoto, M. and Kawaguchi, M. (1973) Pharmacological actions of *Valeriana officinalis* var. *latifolia*. *Yakugaku Zasshi* **93**, 599–606.

Takamura, K., Kawaguchi, M. and Nabata, H. (1975a) Preparation and pharmacological screening of kessoglycol derivatives. *Yakugaku Zasshi* **95**, 1198–1204.

Takamura, K., Nabata, H. and Kawaguchi, M. (1975b) Pharmacological action of kessoglycol 8–monoacetate. *Yakugaku Zasshi* **95**, 1205–1209.

Thies, P.W. (1966) Uber die Wirkstoffe des Baldarins 2: Zur Konstitution der Isovaleriansaureester Valepotriat, Acetoxyvalepotriat und Dihydrovalepotriat. *Tetrahedron Letters*, 1163–1170.

Torsell, K. and Wahlberg, K (1969) Isolation, structure and synthesis of alkaloids from *Valeriana officinalis* L. *Acta Chem. Scand.* **21**, 53–62.

Tortarolo, M., Braun, R., Huebner, G.E. and Maurer, H. (1982) In-vitro effects of epoxide-bearing valepotriates on mouse early hematopoetic progenitor cells and human T-lymphocytes. *Archives of Toxicology* **51**, 37–42.

Tucker, A.O. and Tucker, S.S. (1988) Catnip and the catnip response. *Econ. Bot.* **42**, 214–231.

Veith, J., Schneider, G., Lemmer, B. and Willems, M. (1986) Einfluss einiger Abbauprodukte von Valepotriaten auf die Motilitat Licht-Dunkel Synchronisieter Mause. *Planta Medica* 179–183.

Wagner, H. and Jurcic, K. (1979) Uber die spasmolytische Wirkung des Baldrians. *Planta Medica* **37**, 84–86.

Wagner, H. and Jurcic K. (1980) *In-vitro* und *in-vivo* Metabolismus von ^{14}C-Didrovaltrate. *Planta Medica* **38**, 366–376.

Wagner, H., Jurcic, K. and Schaette, R. (1980) Vergleichende Untersuchungen uber die sedierende Wirkung von Baldrianextrakten, Valepotriaten und ihren Abbauprodukten. *Planta Medica* **38**, 358–365.

Willey, L.B., Mady, S.P., Cobaugh, D.J. and Wax, P.M. (1995) Valerian overdose: a case report. *Vet. Human Toxicology* **37**, 364–65.

Zhang, B.H., Meng, H.P., Want, T., Dai, Y.C. and Shen, J. (1982) Effects of *Valeriana officinalis* L. extract on the cardiovascular system. *Yao Hseuh Pao* **17**, 382–384. (*International Pharmaceutical Abstracts* **20**, 11354).

4. CULTIVATION OF VALERIAN

JENŐ BERNÁTH

Department of Medicinal Plant Production, University of Horticulture and Food Industry, 1114 BUDAPEST, Villányi út 29/31, H-1502 Hungary.

CONTENTS

SELECTION

Self-fertilisation of the species
Crossing experiments
Cultivars and cultivated populations

CULTIVATION

Location
Preparation of the land
Sowing
Planting
Care of plants
Harvesting
Seed production

INTRODUCTION

Valerian (*Valeriana officinalis* and its related species) is cultivated in many countries of
Europe and Asia. It is cultivated in relatively large areas of Russia, Ukraine, Poland,
Bulgaria, Romania, Hungary, Belgium, France. Other *Valeriana* species (e.g. *V. wallichii*,
V. fauriei) are cultivated as well in India, Japan and other countries. In this chapter the
selection of material for cultivation, the factors influencing optimum growth and the
cultivation methods used in the temperate zone are discussed in more detail.

PHYSIO-ECOLOGY OF VALERIANA OFFICINALIS

Light

Photoregulation

The importance of light in germination of valerian has received some attention. Sváb
(1978) showed that germination is more intensive under light and at about 20°C
temperature. Similar phenomenon has been proved by the data of Dagyte and Morkunas
(1968). They concluded that, under light and room temperature (16 − 18 °C), the
germination power of seeds was as high as 34 % and 44 %. In darkness germination
was only 22.2 % and the total number of germinated seeds less than 30 %. The
importance of photoregulation in germination processes was studied by Berbec (1970)
as well. He confirmed that, while the light increases the germination percent , it has
only a slight effect on the length of period needed for the appearance of the first germ.
However the germination is regulated by the interaction of light and temperature. As is
obvious from the data of Fig. 1., the effectiveness of light is dependent on temperature
conditions.

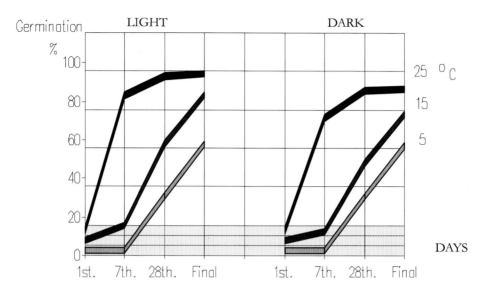

Figure 1 Interaction of light and temperature on the germination of valerian (based on the data of Berbec, 1970)

The effectiveness of light proved to be an optimum one at 25 °C, when the germination power (measured on 7th day) was higher by 20 % and the number of seeds which had germinated by the end of experiment was greater by 10–11 percent, compared with the dark control. The effectiveness of large-energy illumination and long wavelength (328 A°) laser treatment on seed germination was studied by Kuznechova *et al.* (1984). On the basis of their results the phytochrom system of the seeds was stimulated by 10 minutes treatment of both the large-energy illumination (in 67000 mV/cm^2 dose) and laser treatment (0.3 mV/cm^2) and the germination power and the percentage of germinated seeds became higher. The further development of plants was effected by these treatments as well, which were manifested in acceleration of development and higher biomass production.

Effect of light on growth and development

The distribution of valerian species shows a large ecological diversity. It was proved by Voroshilov (1959) who analysed the distribution of *V. exaltata* and *V. collina* species in both Europe and Asia. These species occur under diverse ecological conditions and could be the elements of different plant systems. They can grow inside the forests under low (3-5 Klux) illumination, in margins and clearings in forests and in sunny places as well, where direct illumination can reach as high as 60–80 Klux intensity. It was emphasised by Heeger (1956) that the adaptation of valerian species to the different light conditions is due to the modification of the leaf surface. This modification of leaf surface is going in parallel with structural changes of plant organs.

Effect of secondary metabolite production

The correlation between light and the accumulation level and composition of secondary metabolites has been investigated in the case of valerian. The comparison of plant samples of the same origin taken from natural habitat and cultivated population may show some differences. Some results were achieved by Sárkány and Baranyai (1958) in this respect. It seemed to be an overall phenomenon that the essential oil content of individuals taken from natural habitats was higher. The ecological parameters of natural habitat and the agrarian condition may differ in many aspects, including that of the light intensity.

Temperature

Thermoregulation

As was mentioned above, the temperature may play an important role in regulation of germination processes. The stimulation of light is obvious at the defined temperature regime only, at about 25 °C values. It has been shown by more detailed investigations by Berbec (1970) that the effect of changing day/night temperature is a more characteristic one compared with the constant values. The highest percentage of germination was obtained in his experiments at 25/15 °C day/night rhythm. The germination was relatively good at low temperature values too, if different day/night regimes (15/5 °C) were applied. However, the germination power of the seeds proved to be dependent on the temperature conditions existing previously at the time of seed setting and ripening. It was proved in the long term investigations (Table 1) that a cold and humid climate promoted the vitality of seeds in this phenophase. The seeds harvested in this climate showed high germination power, up to 90 % values. As it is obvious from the data too, the seeds harvested in an optimal year (1962) have a long-lasting biological vitality. The continuous loss of germination power in this seed material is retarded compared with the commercial seed samples. Even after three years using this biological valuable seed material, a relatively good germination of 47 % can be obtained.

Table 1   Effect of climatic conditions existing before ripening on germination percent and life span of seeds (based on the data of Berbec, 1970)

| | Climatic conditions (May-June) | | | Germination % at different time | | | |
	Temp. °C	Precipit. mm	Number of rainy days	After harvest	1 year old	2nd summer	3rd summer
1959	16.9	248.2	27	82.2	63.0	30.2	7.0
1960	15.7	423.7	34	68.0	43.2	18.0	3.8
1961	15.6	147.7	40	78.0	50.0	34.2	9.0
1962	13.7	317.7	55	90.2	89.2	76.4	47.0
1963	17.6	96.6	33	72.0	48.0	36.2	4.6

Effect of temperature on growth and development

It is stated by several authors that the development of valerian is accelerated by dry and hot climatic conditions and that early flowering is followed by quick ripening. On the basis of observation of Dolja (1986), if the temperature increases from 15.7°C to 20.2–21.9 °C before flowering, the time period needed for flowering of *V. collina* and *V. exaltata* becomes much shorter. Similar results were obtained by Konon (1978). He observed that the flowers are open for 4–5 days under cold and rainy weather, while for 2–3 days under dry and hot conditions. This means that the opening of one flower head may continue over 25 to 35 days. The shorter flowering period - effected by hot and dry conditions - has an adverse effect on flowering resulting in restricted vitality of pollens and a decrease in germination percentage of the seeds afterwards.

Effect of temperature on secondary metabolite production

It was stated by Heeger (1956) that there is a correlation between the essential oil content of valerian root and the temperature values of the vegetation period. In Germany the moderate climate seems to be the optimum one: both the warm and the cold weather has an adverse effect on accumulation of essential oils.

Water requirement

Effect of water on growth, development and production

It is accepted generally that valerian belongs to the group of species with high water requirement (Heeger, 1956; Sváb,1978).Taking into consideration the geographical and coenological distribution of the species, much more diversity in the water requirement is expected. Hungarian experiences agree with the data of Auster and Schafer (1958) who showed that *V. exaltata* and *V. sambucifolia* require more humid conditions, while *V. collina* prefers warmer habitats, and grows well under relatively dry forests. In *V. exaltata* and *V. collina* the presence of intraspecific adaptation was demonstrated by Corsi *et al.* (1984). Individuals of the same species were compared, which were gathered from Alpine (Apuan) and Mediterranean regions (Trieste). The characteristic features of the adaptation processes to the dry climate as well as to the humid conditions were described. The individuals grown on the dry karst form thick leaves with two layers of palisade parenchyma. In contrast the Alpine individuals have a thin leaves with small epidermal cells and one parenchymatic layer. In parallel with this modification, the vascular bundles of the stem show a more compact character forming a continuous ring with increased amount of metaxylem when the plant is grown on karst. The adaptation ability of individuals to the high water supply was justified in hydroponic experiments by Gzurjan and Manashjan (1980). The plants affected by continuous water supply formed a large amount of stomata on the leaf surface and the parenchyma cells become large and the amount of the vascular bundles increased in both stems and roots. An optimum water supply - as was proved in hydroponic experiments of Babahanjan (1979) - causes an increase in the development of plants and has a favourable effect on dry matter production. His results showed that the development of plants is quicker by 10 days and the dry-matter production is higher about 3 times if the water supply is optimum.

The importance of water supply on biomass production was shown in plot experiments by Berbec (1965). The plants were grown in plots on 30, 45, 60 and 75 % water saturation level. However there were large differences in the development of the plants effected by water supply and their production was evaluated only at the end of vegetative growth. The optimum biomass of roots were measured at 45 and 60 % saturation levels in all the three vegetation cycles. Both the higher an lower saturation levels had an adverse effect on production. However the low water saturation (30 per cent) decreased the root production in every vegetation cycle universally. The advantage of irrigation was proved under Bulgarian conditions cultivating 'Shipka' cv. as well (Nedkov and Slavov, 1989). The highest yield was achieved by them, over 3520 kg dry roots/ha, when the field moisture capacity was maintained at 90 % continuously.

Effect of water secondary metabolite production

In was proved by the experiment of Corsi *et al.* (1984) that plants grown under different ecological conditions, parallel with tissue modifications, accumulate different amount of special compounds. In the case of *V. collina* the characteristically wide spectrum of essential oil can be measured in roots taken from Alpine region. Under dry conditions, especially in plants grown on karst, the spectrum of essential oil components decreases and 5 of the 25 characteristic essential components present in the Alpine region plants do not accumulate. However, the profiles of the components accumulated under both ecological conditions are not the same. Whilst in the oil of Alpine region plants bornyl acetate (21.7 %) and isobornyl acetate dominate, bornyl isovaltrate is the main component in the oil from the population grown on the karst.

In the systematic experiments the favourable effect of the continuous water supply did not result in increased essential oil accumulation. On the contrary, in the hydroponic experiments of Babahanjan (1979), the essential oil content of the root decreased from 0.3 % to 0.19% as an effect of continuous water supply. The experiments of Berbec (1965) led to the same result. As it is shown in Table 2 the increasement of water saturation level from 30% up to 75 % had on adverse effect on essential oil accumulation.

The essential oil content decreased from 0.77 % to 0.57 % and the biological activity of the root extract decreased as well. The relative value of the biological activity decreased from 408 to 298. Lewkowicz-Mosiej (1984) established the same correlation between water and essential oil accumulation; the investigation was also extended to the analysis of valepotriates. Increasing the water saturation of the soil led to a decrease in the

Table 2 Effect of water supply on drug production of valerian and its content of active constituents (after Berbec, 1965)

Water saturation of soil %	Root production (g)	Essential oil (%)	Essential oil production (mg)	Biological activity (relative value)
30	19.2	0.77	0.15	408
45	27.2	0.72	0.19	391
60	28.8	0.69	0.19	333
75	25.8	0.57	0.14	298

essential oil content of roots from 0.73 to 0.68 % while the amount of valepotriates changed only slightly from 0.63 % to 0.60 %.

In contrast the fatty oil accumulation in the seed can be stimulated by water supply. Dolja (1986) established that the characteristic fatty oil content of a population (20.0 – 23.7 %) can be increased by irrigation 24.2 – 28.1 %. However these quantitative changes are connected with compositional modifications: as a result of the irrigation the accumulation of palmitin, stearin, olein and linolene is increased, while the amount and ratio of erucic acid is much higher in seeds ripening on non irrigated plants.

Soil conditions

Effect of soil on growth, development and production

It was generalised by Heeger (1956) that the heavy clay soils are unsuitable for *V.officinalis*. In harmony with this statement the medium loam or sandy loam soils, which have a medium humus content, are suggested for cultivation by Sváb (1978). The effect of loam and light sandy soils was compared in three years of plot experiments of Berbec (1965a). The advantage of a loam soil was obvious, in which the plants developed more intensively, the number of leaves increased and the plant became 25–30 cm higher. The higher productivity of these plants was shown by the changes in the root formation and production. The number of primary roots was higher in loam soil and the total biomass of roots increased about 100 per cent. The importance of soil characteristics on production of valerian was also shown by Bernáth *et al.* (1973). Comparing different soil types of Hungary, the advantage of the sandy (Örbottyán) and loam soils (Balatonfenyves) were demonstrated in production of root biomass (Table 3.).

Table 3   Effect of some Hungarian soil type on the root production of valerian in average of two vegetation cycles 1971-1972 - (Bernáth et al. 1973)

Soil type	Total biomass/ plant (g)	Root production		Root/ shoot ratio
		Root mass (g)	Ratio of root thinner as 1.5 mm (%)	
Sandy soil - Örbottyán	279.4	151.6	36.3	1.186
Soil rich in clay- Budakalász	212.4	91.5	8.4	0.756
Soil rich in clay- Szarvas	313.9	137.3	6.0	0.743
Loam soil- Balatonfenyves	396.2	156.6	20.9	0.653

Örbottyán - pH in water 8.0, total salt 0 %, $CaCO_3$ 0.2 %, hardness number 27, humus 0.87%, total nitrogen 0.051 %.

Budakalász - pH in water 8.2, total salt 0.02%, $CaCO_3$ 7.7%, hardness number 37, humus 2.33 %, total nitrogen 0.137 %.

Szarvas - pH in water 8.0, total salt 0.04%, $CaCO_3$ 1.0 % , hardness number 44, humus 4.12 %, total nitrogen 0.242 %.

Balatonfenyves - pH in water 8.2, total salt 0.04%, $CaCO_3$ 10.7 %, hardness number 49, humus 13.72 %, total nitrogen 0.805%.

The root production of individuals was 151.6 g in sandy soil, 156.6 g in loam, while only 91.5 – 137.3 g in the other soil types are rich in clay components. The morphological character of the root was affected by the soil type as well. In the sandy and loam soil (Örbottyán, Balatonfenyves) the relative mass of thin roots increased, taking about 20.9 – 36.3 per cent of total biomass of roots. In the clay soils the ratio of the less valuable parts of roots decreased showing 6.0 – 8.4 % only. The large amount of thin roots and the high root/shoot ratio in sandy soils, as well as in loam (with low mineralization ability) reflects on the good adaptation of the species which copes with a lack of nutrients by intensive differentiation of thin roots.

Effect of soil on secondary metabolite production

In accordance with practical observations it was reported by Berbec (1965a) that the essential oil content of the roots and the biological activity of the extracts made from the roots are dependent on the growing conditions, especially on the soil type. 1.08 % essential oil level was measured in roots taken from loam soil, while 0.87 % in the sandy-clay only. In contrast the biological activity of the extract was higher in the roots grown in the latter soil type. The effect of soil on accumulation of special compounds was justified by plot experiments (Bernáth et al. 1975).

Evaluating the data shown in Fig. 2 the maximum values of essential oil accumulation were observed in plants grown in different soil types, especially in clay (Szarvas) and sandy (Örbottyán) soils. From the point of view of the root tissue structure the increasing root surface would result in a potentially higher accumulation but the values show contradictory results as the essential oil accumulation is independent from the ratio of the rootlets being thicker then 1.5 mm. It does mean that there is no correlation between

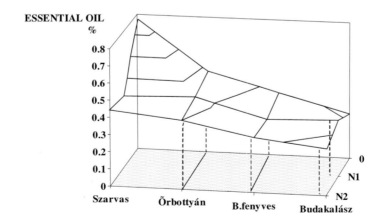

Figure 2   The effect of Hungarian soil types on essential oil accumulation of valerian, based on the plot experiment of Bernáth et al. (1975)

the actual essential oil content and the number of the cells which are able to accumulate the oil. In the case of the valepotriates another type of correlation was observed. The accumulation of these substances was more restricted in the soil Szarvas, which was characterised by presence of high amount of nutrients.

Nutrition

Effect of nutrition on growth, development and production

It was emphasised by Golcz and Kowalewski in 1958 that there are many contradictory results relating of the nutrient requirement of valerian. From practical purposes the selection of the soils of high humus and nutrient content are proposed by many authors (Heeger, 1956, Sváb, 1978). The reason for the contradictory results is that the efficacy of the nutrient supply may depend on many factors. The results of Lewkowicz -Mosiej (1984) demonstrated that the water saturation level may determine the effect of nutrient fertilisation: at 80 % water saturation level the activity is significant while at 40 % water saturation level it is only slight from the same dose of nutrients. The exact reaction profile of plants to the different nutrients (nitrogen, phosphorous and potassium) can be characterised in plot experiments (Bernáth *et al.* 1973). The increase in the degree of nitrogen supply from 10 mg N pot^{-1}week^{-1} to 800 mg N pot^{-1}week^{-1} doses resulted in an optimum type growth in fresh-mass production of plants (Fig. 3.).

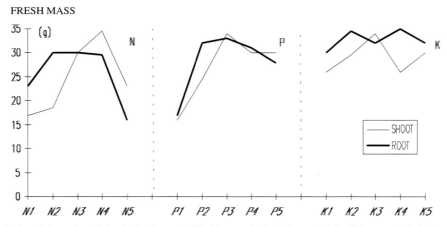

N supply (mg N pot^{-1}week^{-1}): (N1)10 mg, (N2) 50 mg, (N3) 200 mg, (N4) 400 mg, (N5) 800 mg,
P supply (mg P pot^{-1}week^{-1}): (P1)5 mg, (P2) 25 mg, (P3) 100 mg, (P4) 200 mg, (P5) 400 mg
K supply (mg K pot^{-1}week^{-1}): (K1)5 mg, (K2) 25 mg, (K3) 100 mg, (K4) 200 mg, (K5) 400 mg

Figure 3   The effect of nitrogen, phosphorous and potassium supply on the shoot and root production of valerian, based on the plot experiment of Bernáth *et al.* (1973)

However, while the optimum point of shoot production was measured at N4 dose, the amount of roots reached the maximum value at lower nutrition supply (N2-N3). The higher nitrogen requirements of the shoot and and relatively smaller requirements of root development are well demonstrated by the changes of their ratio. Parallel with the increasement of nitrogen supply the relative value of the root decreases from 1.356 to 0.696. The structural characters of the root change at the same time. As an effect of the increasement of the nitrogen doses the thickness of the roots increases. The thickness of the roots can be characterised by the dry-mass production of the roots calculated to the unit surface area of root. Using this calculation method the 5.25 g cm^{-2} value measured at low nutrition increases up to 16.08 g cm^{-2} at high dose of nitrogen.

Changes of phosphorus supply result in changes similar to those described for nitrogen application above. There is only a difference in the rate of shoot and root production, which shows no characteristic changes.

The potassium supply has a regulatory effect on the structure of the root similar to nitrogen. An increase results in a restriction in leaf-surface and parallel increasement of root thickness.

The interaction of nitrogen and phosphorus is demonstrated in Fig. 4a-c. Unambiguous interaction was found in the case of biomass production of root (a). Otherwise the shoot production is mainly regulated by nitrogen supply (b) and the structural changes of the root, characterise by its dry-mass production calculated to the unit surface area (c) are regulated by both nitrogen and phosphorus.

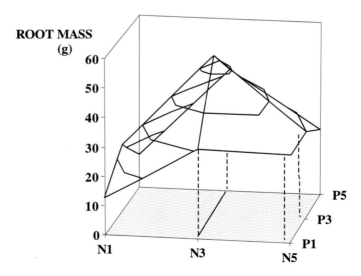

Figure 4a Interaction of nitrogen, phosphorus and potassium supply on the root production of valerian, based on the plot experiment of Bernáth *et al.* (1973).

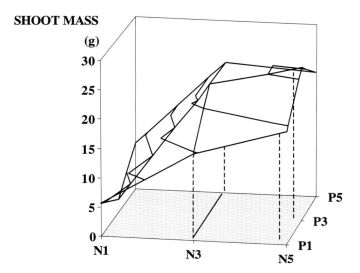

Figure 4b   Interaction of nitrogen, phosphorus and potassium supply on the shoot production of valerian, based on the plot experiment of Bernáth *et al.* (1973).

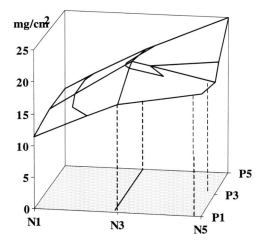

Figure 4c   Interaction of nitrogen, phosphorus and potassium supply, based on the plot experiment of Bernáth *et al.* (1973) on structural changes of the root, characterised by dry-mass production calculated to the unit surface area of root.

There is some information on the calcium requirement of the species. It was mentioned by Heeger (1956) that valerian grows well in lime-soils rich in calcium. This phenomenon was proved experimentally by Berbec (1965a). However it was clear from the results that the effectiveness of calcium dosage is dependent on the chemical form of calcium as well as on the soil-type chosen: CaO in loam and $CaCO_3$ in sandy-clay accelerated the growth of roots. Calcium fertilisation also improved the root quality, the utilizable part of rootlets increasing by 75.2 % to 78.8 %.

Effect of nutrients on secondary metabolite production

The real connection between the accumulation of special compounds and the nutrient supply is hardly known (Golcz and Kowalewski, 1958). The reason for the contradictory results is that the majority of the authors make no distinction between the changes of the accumulation level and those of production. Lewkowicz - Mosiej (1984) stated that the application of nitrogen, phosphorous and potassium may decrease the accumulation level of the essential oils. However, these changes do not affect the essential oil production of individuals, which is due to the acceleration of biomass production through use of fertilisers. The application of fertilisers also increases the production of valepotriates without any changes in accumulation level. In plot experiments the essential oil accumulation, which was affected by nutrition supply, showed an optimum curve (Bernáth *et al.* 1975).

As shown on Fig. 5., there is no correlation between the accumulation level of the essential oils and the structural changes of the root morphology. Increasing the nutrient

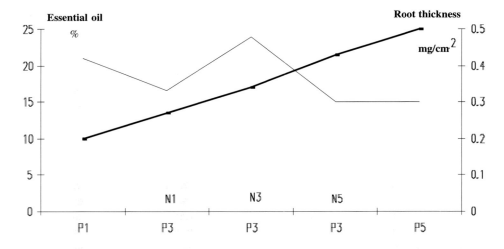

Figure 5   Effect of nitrogen and phosphorous supply on root structure and essential oil accumulation of valerian, based on the plot experiment of Bernáth *et al.* (1975)

supply increases the thickness of the roots. In thin roots the relative amount of the hypodermic layer in which the accumulation of essential oils takes place is relatively high. In contrast the ratio of hypodermic cells in thick roots is smaller. However, the accumulation of essential oil is independent of these structural changes. In the case of valepotriates some kind of optimum effect was proved by the investigations of Bosetto *et al.* (1987). Parallel culture of plants in containers and in a hydroponic system showed that medium nutrient doses, 150 kg N, 75 kg P_2O_5 and 75 kg K_2O/ha resulted in the highest root and valepotriate production.

GENETICS AND SELECTION OF VALERIAN

Genetic variability in natural populations

It has been proved by several investigations carried out in natural habitats that the populations of valerian can be characterised by high variability in morphology and in the production of constituents (Titz and Titz,1982: Titz *et al.* 1983). This variability is much greater if individuals of the same species are compared taken from deverse geographical sites and ecological conditions. The reason for the morphological variability of the genus was explained by Czabajska (1964) by the differences in chromosome number which were found in Poland. The majority of the individuals of the local *V. officinalis* var. *latifolia* population belonged to the diploid (2n=14) form, which was characterised by high variability in leaf morphology, stem height and time of blossom. On the basis of the more detailed investigations of 27 natural populations the occurrence of a tetraploid form (2n=28) was demonstrated. The leaves of the tetraploid individuals were much wider compared to the diploid ones. There were also differences in the size of the pollen grains, the diameter being about 46.6 μm in tetraploid plants and 42.3 μm in the case of the diploid form. The existence of individuals showing different chromosomatic features was confirmed by the investigations of Goldblatt and Johnson (1990). Analysing *V. officinalis* and *V. officinalis* subsp. *officinalis* populations the chromosome number was determined to be a diploid one (2n=14), while the populations of *V. officinalis* subsp *collina* proved to belong to the tetraploid group.

The diversity of chromosome number was also shown in the case of Bulgarian populations (Evstatieva *et al.* 1993). *V. officinalis* was found to be a part of an extremely polymorphic polyploid complex. On the basis of morphological, cytological and phytochemical investigations the two subspecies '*officinalis*' and '*collina*' were each divided into diploid (2n=14) and tetraploid (2n= 28) cytotypes.

The close correlation between the number of chromosome and the variability of the populations was described by Noller (1989). Comparing populations of German, Polish and Hungarian origin this correlation was proved in relation to both morphological and chemical characters (Table 4.).

In contrast to earlier investigations the existence of the octoploid (2n=56) was shown which reflects the possible autopoidity of the genus. From the results the populations, determined as *V. exaltata*, *V. collina* and *V. procurrens,* showed a high morphological and chemical diversity in relation of their chromosome number. The population with higher polyploidy formed much larger and wider leaves, with more and larger stomata on them.

Table 4 Morphological and chemical characters of populations of German, Polish and Hungarian origin in relation of their chromosome number (based on data of Noller 1989)

Origin of the population	Taxa	Number of chromosomes	Leaf length cm	Leaf width cm	Length of stomata mm	Size of seed mm	Mass root g/plant	Active agent mg/100g HVA	AVA	VA	VAL	Essential oil ml%
Sauerland	'exaltata'	14	8.0	3.5	46	2.4x1.2	23.1	52.5	12.9	21.8	5.0	-
Berchtesgaden	'collina'	28	5.5	3.1	48	2.4x1.2	66.9	63.7	93.5	32.2	8.7	1.04
Reiskirchen	'collina'	28	5.4	3.0	46	2.9x1.2	66.9	0.0	70.8	20.6	2.5	0.89
Rödgen	'collina'	28	8.5	4.0	48	2.7x1.2	21.6	16.4	18.1	5.0	0.0	-
Taunus	'procurrens'	56	13.3	7.8	72	3.6x1.8	118.1	5.9	74.9	161.2	27.2	-
Wieseck	'procurrens'	56	9.5	6.7	64	3.8x1.8	79.8	2.4	87.2	138.9	21.6	0.45
Heuchelheim	'procurrens'	56	5.3	3.3	68	3.4x1.6	16.8	106.1	18.7	171.8	22.9	-
Hungarian	'procurrens'	56	13.0	6.6	64	3.9x1.1	26.9	163.6	138.9	264.7	41.4	-

HVA= Hydroxy-valerenic acid, AVA=Acetoxy- valerenic acic, VA= Valerenic acid, VAL=Valernale

The size of the seeds increased as well. Considerable changes were also found in the accumulation of the special plant compounds. The higher chromosome number resulted in higher accumulations of valepotriates and valerenic acid; the relation of the valerenic acid content to chromosone number seems to be very strict. While the valerenic acid content in the root of the diploid and tetraploid forms was as low as 5–32 mg%, in individuals of octoploid plants an accumulation level of 138.9–264.7 mg% was found.

Genotype-phenotype interaction

It was emphasised by Noller (1989) that, in evaluating the variability of the natural populations of valerian, the modification effect of the ecological conditions has to be considered. Standardisation of the genetical and ecological interactions is one of the preconditions to the selection of new cultivars of stable and high productivity. Some aspects of interaction were proved by Noller (1989) comparing two cultivars 'Polka' and 'Erfurter Breitblattiger Baldrian' under phytotron condition using 5 different temperature programs. Comparing the two cultivars in a short vegetation cycle there were only 19–46% differences in accumulation of valerenic acid. When the vegetation cycle was extended by three month the differences between cultivars increased up to 52–94%. However these differences could only be generated under higher temperature regimes. The accumulation of valerenic acid decreased in both cultivars equally when they were grown under low temperature regimes.

Morphological characters and content of active ingredients

The determination of any connection between the morphological characters and the levels of production of secondary compounds in valerian would be a great help in practical selection of plants. Although the existence of this type of correlation between taxa of diverse polyploidy has been shown by Noller (1989), no evidence was submitted with regard to populations having identical chromosome number. On the basis of investigations had been extended to three vegetation cycles, only slight and year dependent correlation were registered (Table 5.).

Table 5 Correlation ('r' values) of morphological and chemical chracters of plants (based on the data of Noller 1989)

Character	Year	Root production	Plant height	Size of leaves	Number of leaflets
Valerenic acid	1986	0.13	-	0.06	- 0.36
	1987	0.07	0.03	- 0.23	- 0.24
	1988	0.01	0.27	- 0.37	- 0.32
Valerenale	1986	0.11	-	- 0.14	- 0.32
	1987	0.26	0.24	0.23	- 0.24
	1988	- 0.04	0.36	0.34	- 0.32
Essential oil %	1986	- 0.06	-	- 0.08	0.17
	1987	0.18	- 0.01	0.096	- 0.09
	1988	0.09	- 0.29	0.43	- 0.17

Table 6   Correlation ('r' values) between chemical compound accumulated in the root (based on the data of Noller, 1989)

Chemical compound	Year	Hydroxy-VA	Acetoxy-VA	Valeranale	Essential oil %
Valerenic acid (VA)	1987	0.65	0.61	0.64	0.03
	1988	0.00	0.33	0.75	- 0.19
Hydroxy-VA (HVA)	1987	-	0.76	0.76	0.08
	1988	-	0.18	- 0.08	0.06
Acetoxy-VA (AVA)	1987	-	-	0.61	0.23
	1988	-	-	0.33	0.09
Valerenale (VAL)	1987	-	-	-	0.17
	1988	-	-	-	- 0.22

Evaluating the data which were shown in the Table 5., it is obvious that the "r" values are much below the values expected and the variability affected by the condition of the year is rather high. Only the existence of a negative correlation between the number of leaflets and accumulation of valerenic acid and valerenal are worth mentioning. The absence of the correlation between morphological characters and level of active constituent is emphasised by other publications approaching the subject from ecological point of view (Bernáth et al. 1975).

The existence of correlation between the compounds has been proven by mathematical analysis (Table 6.).

These correlations between secondary compounds have great importance from a theoretical and practical point of view and is not neglected by the considerable influence of the year which was chosen for the investigation. Based on the existing correlation the parallel increase of valerenic acid (VA), valerenal (VAL) and hydroxy valerenic acid (HVA) is expected by the systematic selection. A slight negative correlation was found between the accumulation of valerenic acid and the accumulation of the essential oil. The absence of correlation fort valepotriates and essential oils has been mentioned by other authors (Bernáth et al. 1975).

SELECTION

Self fertilisation of the species

The advantage of using self-fertilising lines was mentioned at first by Czabajska (1964) and Shugaeva (1979). Some years later the result of self-fertilisation using individuals of 'Kordiola' cultivar was reported by Konon and Novikova (1981). When the vitality of the inbred lines and the original population were compared the negative effect of self fertilisation became obvious. While the seed production of the control plants was as high as 2–15 g, many of the self fertilised plants became sterile and the average seed production of individuals decreased to 0.20-0.41 g. At the same time the germination power and the germination percent of the seeds, which were collected from self fertilised plants was also reduced. The plants which were grown from these seeds showed many negative characters associated with self-fertilisation; deformations were observed in reproductive organs, male sterility appeared and biomass production was reduced. However, from the point of view of plant breeders the increase of both morphological and chemical variability was considered as a positive phenomenon. The increase in variability can be characterised by the values of the variation coefficients calculated for these parameters. The coefficient of variation of plant height increased by up to 43–68% and that of the root production by up to 44–46 % in the I1 generation. In respect of the essential oil the individual values of the original population were near the mean (0.34 %) with low variability. After self-fertilisation the quantitative distribution of the essential oils became heterogeneous and values as high as 0.47–0.54 % were recorded. These results emphasise the importance of self fertilisation in the selection of new varieties.

Crossing experiments

The importance of plasmatic inheritability was proved by Noller (1989) in reciprocal crossing experiments. The presence of plasmatic inheritance was justified in relation to both morphological-production biological and chemical properties. In the model experiment using line number 12 for crossing, the inheritability of quantitative characters in F1 generation was determined whether the line was used as a mother plant or as a pollen donor. When it was used as a pollen donor the valerenic acid content in F1 generation was 90 mg% or higher in the majority of individuals (85%). In contrast if the line was used as a mother plant the ratio of individuals accumulating high amount of valerenic acid decreased to below 42 %.

It was concluded by Noller (1989) that, taking into consideration the polyploid character of the species, the biometrical analysis of inheritability encounters many difficulties. Based on his experiments it was concluded that, before crossing, at least 4–6 inbred generations from diploid lines are needed, while 16–24 generations are needed from tetraploid plants. To clear up the additive genetic variance the application of the "Chain block design" strategy was proposed by Noller and Vömel (1989). Additive gene effect on 0.02 significance was determined by the German research group in relation of the VA, VAL accumulation level as well as in production of VA and VAL compounds (Table 7.).

Cultivars and cultivated populations

The efficacy of the selection of valerian is moderate compared with the large agricultural crops, even compared with the results that have been achieved in the case of some other medicinal plants. The selection process has been hindered by many factors. The

Table 7 Additive gene effect in %, measured in crossing experiments by Noller (1989)

Chemical character	Accumulation level (mg/100 g)	Production (mg/plant)
Valerenic acid (VA)	91.0	88.7
Valerenale (VAL)	84.3	88.2
	Significance level 0.001 %	

variability of the chromosome number (2n=14, 2n=28, 2n=56) causes difficulties in the stabilisation of the characters required. The goal of the selection should be the stabilisation of different production-biological and chemical characters, which may correlate to each other negatively. From a selection point of view there are other difficulties, which are generated by the advancement of chemical and biological science discovering new compounds continuously. So the goal of the selection has to be modified from time to time. Taking into consideration the above mentioned facts, four periods in the selection of valerian can be distinguished.

The first period of the selection of valerian is dated back to the early 1930s. The selection of the native populations was started by evaluating their morphological and production biological characters, especially their root productivity. The accumulation of the essential oil was the only chemical character which was taken into consideration. As a result two German cultivars were registrated in 1932. Their names 'Schmalblattriger Baldrian' and 'Breitblattriger Baldrian' reflects the morphological approach that was followed in the course of their selection. By the improvement of these cultivars three new materials were developed in Germany and registered in 1942. The names of these cultivars were; 'Erfurter Baldrian', 'Oderland Baldrian' and 'Oberlausitzer Baldrian'. As a result of the further selection tetraploid and octoploid cultivars were developed and registered in 1955. The 'Frankfurter Schmalblatriger Oderland Baldrian' and the 'Oberlausitzer Schmalblatriger Baldrian' surpass the earlier cultivars in both root and essential oil production.

The second period of the selection of valerian was connected with the introduction of "population analysis" begun in the middle of 50s (Eisenhuth 1956). The result of this period was the registration of the cultivar `Merkator` in Germany. Its root and essential oil production were about 20 % higher than earlier cultivars. However, this cultivar had to be propagated vegetatively which was a great disadvantage to its commercialisation. The results achieved in selection of valerian by Czabajska (1958) at this time should also be mentioned.

The third period of the selection of valerian started in the early 1970s. The stabilisation of the production was chosen as the main aim of the selection. The weakness of the former cultivars became obvious in the course of cultivation. The yield and the quality of the root showed much more variability than was tolerable to the farmers. The selection of the resistant materials with high essential oil production had been started. At that time the isolation of the biologically active valepotriates had a remarkable effect on further selection process. Well-known varieties introduced in this period were: 'Anthos' (East-Germany), 'Samokov' (Bulgarian) and 'Polka' (Polish).

The present phase of selection of valerian can be characterised by the application of the new genetical knowledge that has been achieved recently. In the process of selection the traditional and new methods of plant breeding are applied in parallel (Noller 1989). It is emphasised by the majority of publications (Kempf 1986, Vömel and Hölzl 1989) that the idea of the selection process has to be modified to give more attention to the chemical constituents, especxially for essential oil, valepotriates and valerenic acid. Furthermore the parallel checking of biomass production as well as the accumulation of many special compounds (essential oil, valerenic acid, valeranone, cyclopentan-sesquiterpenes) are suggested by working group of Bos et al. (1986).

CULTIVATION

Location

Valerian can be cultivated successfully in almost any area of Europe and many parts of Asia, where the natural precipitation is about 600–700 mm/year. For germination, light and a moderately high temperature (about 20°C) is also required. Valerian prefers medium loam or sandy loam soils, which have a medium humus content. The root system grows well in chernozem and peat soils, with a high humus content, but these ought to be avoided because of the difficulties arising from the adhesion of soil particles in postharvest processing. The location must be well supplied with moisture or equipped with an irrigation system.

Preparation of the land

Valerian can be successfully cultivated after almost any agricultural crop, because it does not show special requirement with regards to the previous plant (Hornok, 1992). However, crops with perennial roots like lovage, liquorice, peppermint are not the best ones to precede cultivation of valerian.

If planting is done in the autumn after the early harvest of the previous crop medium to deep ploughing is necessary just after harvesting. It is followed by harrowing or cultivator operation combined with fertiliser supply. The field up to the time of the planting must be kept clear from weeds. In the case of the spring planting, deep ploughing is necessary in the previous autumn and the soil has to be prepared for planting in early spring.

Sowing

Valerian can be propagated both by seeds and vegetatively using root segments (Boros 1980). The vegetative propagation method is applied only in the case of varieties which are able to develop large amount of stolons. Vegetative propagation may have another disadvantage. The plants frequently form flower shoots in the first year of cultivation, which appears to restrict root development. The effectiveness of vegetative propagation is also dependent on the ecological properties of the cultivation area. According to Heeger (1956) the optimum yield of root can be obtained at the end of the first vegetation cycle using vegetative propagation in a moist environment, while only after the second one if the climate is dry.

The seeds of valerian can be sown at various times of the year and by different methods. Direct sowing is widely used in the more humid areas (in some districts of Poland and Ukraine). Catizone *et al.* (1986) proposes the direct sowing in autumn using 50 cm row distance and 3 kg/ha seed dosage. Under Yugoslavian conditions (Kisgeci *et al.* 1987) the optimum time of direct sowing seems to be later, in October or in the first half of November, and with higher amount of seed (7-10 kg/ha). Studies by Hotyn *et al.* (1967) revealed that the time and the form of the direct sowing is determined by the local climatic conditions. Evaluating the cultivation trials made in Belorussia, Ukraine and West-Siberian regions the success of the direct sowing is dependent on the ecological conditions preceding wintering. The plants have to reach the development phase of 3–5 leaves before the cold period which is a precondition of successful wintering. A special method is applied when valerian is grown together with a second crop. In Russia barley, rye and oats are often intermixed with valerian. In this case the amount of propagative material of valerian is decreased by about 20–25 %, and the second crop is harvested in spring for fresh biomass.

However in some countries, especially in Hungary, France and Poland nursery sowing in open-air seedling beds has proved to be the most reliable and economic way of propagation (Bernáth 1993, Perrot and Paris 1974, Rumiska 1973).

In the case of nursery sowing the procedure has to be started between the end of June and early August. The seeds should be sown into well-prepared open-air seedling beds on the surface of the soil using 15–20 cm row distances. According to the literature (Catizone *et al.* 1986, Kisgeci *et al.* 1987) 1–2 g of seeds is required for 1 m² area. The seeds are covered afterwards with compost very carefully using no more than 1 mm layer About 0.5–0.7 kg seed is required to grow seedlings for one hectare cultivation area and 500–700 m² nursery surface. One of the preconditions of good development of seedlings is regular irrigation. The nursery should be slightly shadowed in order to reduce the water loss through evaporation until the appearance of the first leaves. The period which is needed for the seedlings to achieve this is about 2.0–2.5 months.

Planting

The seedlings can be planted into the field when they reach the 15–17 cm size at the end of September (Bernáth 1994), or in spring (Perrot and Paris, 1974) using machines. To allow sufficient time for rooting, the plantation time must be about 1–2 months before the first frost occurring in the cultivation area, spring planting should be avoided. If it is necessary, it should be done as early as possible, not later than April. However, using this cultivation form the yield is generally about 20–25 percent less compared to autumn culture. If the plantation is cultivated by hand the optimum distance between rows is 30–35 cm and 20–25 cm between individual plants. In the case of mechanised cultivation, the row distance should be larger, 50 cm at least, with the same placing in the rows. However the row distance may vary from country to country. Large 60–80 cm row distances are proposed by Catizone *et al.* (1986), Perrot *et al.* (1974) and Heeger (1956), while 30–40 cm by Kisgeci *et al.* (1987) and 50 x 25 cm spacing by Racz *et al.* (1984). However on the basis of investigation of Berbec (1968) the optimum spacing is about 40 x 40 cm. The difference in spacing may also depend on the various ecological conditions and the local cultivation methods.

Care of plants

After wintering, the plantation has first to be rolled. This is an important action especially if the root system of the plants has been disarranged as a result of winter frosts. The plantation requires two or three weedings and hoeings. Because of the high water requirement of the species (Berbec 1965) irrigation is necessary under more arid conditions to get satisfactory biomass production. Chemical weed control is also used and, in the case of autumn planting, pre-emergent treatment should be done with 2.5 – 3.5 kg/ha Aresin. If the plantation is left to spring the dosage of Aresin has to be reduced to 2.5 – 3.5 kg/ha. Keeping the plantation clear of weeds afterwards can be achieved with 2.5 –3.5 kg/ha Patoran or Maloran if the plant height is more than 20 cm.

Powdery mildew (*Erysiphe polygoni*) and peronospora (*Peronospora valerianae*) are common fungal pathogens of valerian but the plants can be protected against them by common fungicides.

Harvesting

Valerian which has been planted in the autumn can be harvested in the next year in October. The harvest of the spring culture is done in the year of plantation, very rarely in the next one if the climatic conditions are exceptionally disadvantageous. The optimum time for root harvest is in the second half of October or first half of November. Sometimes the root is left in the ground for wintering just prior to the start of vegetation. This cultivation practice has many disadvantages and may result the decreasement of the accumulation level of active constituents.

Before taking off the roots the above ground parts of the plants should be cut using mobile chaff-cutter machines. After this procedure the roots can be taken up with a plough from which the steering plate has been removed, or by the help of potato harvesting machines. In both cases the effectiveness of the operation is largely dependent on the soil conditions. The roots are usually gathered by hand, put in piles so that the green parts can be removed and the soil particles shaken out.

The roots are washed with water using basket immersion or a water jet. The material should be dried at 40–50 °C temperature without any delay. Many authors propose much lower (25–35 °C) temperature conditions to avoid the loss of active constituents (Boros 1980, Catizone *et al.* 1986). During the drying process and subsequently in storage valerian should be separated from the other drugs, owing to its strong smell.

The drug of valerian must be stored in confined place, to avoid the contamination by cats which are often attracted by its smell.

Depending on the effectiveness of cultivation, the fresh yield of valerian is about 3.5–7.0 t/ha in Hungary, which results about 2.0–4.0 tonnes dry mass. Much higher or smaller yields have been reported by different authors. The yield is dependent on the ecological condition of the cultivation area and on the cultivation method as well. Hotyn *et al.* (1967) reported that, under Russian conditions, 1.2-1.6 tonnes of root can be harvested after an autumn-spring vegetation cycle, while using spring sowing the yield is reduced to about half.

Seed production

Production of seed material for propagation purposes needs special care and technology because of the continuous ripening and dropping of the seeds which starts after flowering.

In the small scale production of seeds the flower heads of the plants are used individually wrapped up in paper sheets or bags at the beginning of flowering. The seeds fall down into the bottom of the wrapping. This method is effective, if the weather conditions are appropriate. Heavy rain makes the paper sheet or bag wet and the seeds start to germinate or rot and strong winds may harm the paper cover resulting in considerable losses. The other disadvantage of this method is that the wrapping and gathering of the seeds is labour-intensive.

In many countries a two stage method is used for producing seeds. In this case the flowering shoots of plants are cut while green and the the first seeds are starting to ripen. The shoots are gathered and stored in a dry and covered place having appropriate air circulation. When the majority of seeds reach the maturity stage they should be thrashed out. The seed yield depends on climatic conditions and on the methods of cultivation and harvesting. In these circumstances the yield can vary in a wide range, between 30 - 200 kg/ha.

REFERENCES

Auster, F., Schafer, J. (1958) *Arzneipflanzen*. 18. *Valeriana officinalis* L. VEB Georg Thieme, Leipzig. 1–44.

Babahanjan, M.A. (1979) Proidvostvo valeriani lekharstvenoi v ushloviah otktitoi gidroponiki. *Izd. Akad. Nauk Arhmanskoi SSR.* **18**, 49–56.

Berbec, S. (1965a) Influence of soil humidity on the growth, yield and quality of the raw material of valerian. *Ann. Univ. Mariae Curie-Sklodowska, Lublin-Polonia*, **20**, 216–231.

Berbec, S. (1965b) Influence of various doses of calcium on the growth, yield and quality of the raw material of valerian. *Ann. Univ. Mariae Curie-Sklodowska, Lublin-Polonia*, **20**, 233–249.

Berbec, S. (1970) Some problems from the biology of seed germination of common valerian (*Valeriana officinalis* L.) *Ann. Univ. Mariae Curie-Sklodowska, Lublin-Polonia*, **25**, 143–152.

Berbec, S. (1968) Influence of spacing and hoeing on the quantity of the yield of valerian (*Valeriana officinalis* L.). *Annales Universitatis Mariae Curie-Sklodowska, Lublin, Polonia*, **23**, 323–338.

Bernáth, J. (1993) *Vadon termő és termesztett gyógynövények*. Mezőgazda, Budapest, 486–489.

Bernáth, J., Földesi, D., Lassányi, Zs. (1973) A tápanyag-ellátottság és a talajtípus hatása a macskagyökérre (*Valeriana officinalis* L.ssp. *collina* (Walr.). I. A növények gyökérszerveződésének, növekedésének, illetve droghozamának változása. *Herba Hung.* **12**, 45–67.

Bernáth, J., Földesi, D., Lassányi, Zs., Zámbó, I. (1975) A tápanyag-ellátottság és a talajtípus hatása a macskagyökérre (*Valeriana officinalis* L.ssp. *collina* (Walr.). II. Az illóolaj és valepotriát tartalom változása. *Herba Hung.* **14**, 37–46.

Boros, G. (1980) *Heil- und Teepflanzen*. Verlag Eugen Ulmer, Stuttgart, 28–29.

Bos, R., Putten, F., Hendriks, H., Mastenbroek, C. (1986) Variation in the essential oil content and composition in individual plants obtained after breeding experiments with *Valeriana officinalis* strains. *Progress in Essential Oil Research*. Walter de Gruyter Co., Berlin-New York, 123–130.

Bosetto, M., Fusi, P., Arfaioli, P. (1987) Indagini sull'influenza della fertilizzazione azoto-fosfo-potassica sulla resa principio attivo (valepotriati) della *Valeriana officinalis* L. *Agrochimica*, **31**, 254–264.

Catizone, P., Marotti, M., Toderi, G., Tétényi. (1986) *Coltivazione delle piante medicinali e aromatiche*. Patron Editore, Bologna, 283–287.

Corsi, A., Lokar, L., Pagni, A.M. (1984) Biological and phytochemical aspects of *Valeriana officinalis*. *Biochem.Syst.Ecol.* **12**, 57–62.

Czabajska, W. (1958) *Valeriana officinalis* L. *Buil.Inst. Roslin, Lecziczych*, **4**, 89–100.

Czabajska, W. (1964) Untersuchungen über Baldrian in Polen. *Pharmazie.* **19**, 468-470.

Dagyte, S.J. Morkunas, A.V. (1968) Some biological properties of the official valerian seeds. *Trudi.Akad. Nauk. Liuthuanian SSR.* B, **2**, 69–74.

Dolja,V.S. (1986) Vlianie poliva na sodhersanie i himicheskih sostav sirnih masel semian nekatorih virashivaemih vidov *Valeriana* L. *Rast.Resursi.* **3**, 348–351.

Eisenhuth, F. (1956) Qualitatsforschung und Leisungsfragen bei *Valeriana officinalis L. Pharmazie,* **11**, 271–286.

Evstatieva, L.N., Handjieva, N.V., Popov, S.S., Pashankov, P.I. (1993) A biosystematic study of *Valeriana officinalis* (Valerianaceae) distributed in Bulgaria. *Plant Systematics and Evolution.* **185**, 167–179.

Golcz, L., Kowalewski, Z. (1958) Wyniki doswiadczen nawozowycha kozlkiem lekarskim (*Valeriana officinalis* L.) *Biul.Inst.Roslin Lecznychych.* **4**, 107–113.

Goldblatt, P., Johnson, D.E. (1990) *Index to plant chromosome numbers,* 1986–1987. Missouri Botanical Garden, 30.

Gzurjan, M.S., Manashjan, K.S. (1980) Osobennosthi stroenia lista i kornia valeriani lekharstevennoi v ushloviah pktiroi gidroponiki. *Izd. Akad. Nauk Arhmanskoi SSR.* **20**, 132–141.

Heeger, E.F. (1956) *Handbuch des Arznei und Gewürzpflanzenbaues.* Deutscher Bauern-Verlag, Berlin, 693–705.

Hornok, L. (1992) *Cultivation and processing of medicinal plants.* Akadémiai Kiadó, Budapest, 170–176.

Hotyn, A.A., Gubanov, I.A., Kondratenko, P.T., Seberstov, V.V. (1967) *Lekarstvennie rasthenia SSSR.* Izdatelstva, 'Kolos', 65–74.

Kempf, I. (1986) *Grundlagen zur Züchtung von Valeriana officinalis L. Baldrian.* Justus-Liebig-Universitat, Gissen, Dissertation, p. 206.

Kisgéci, J., Adamovic, D., Kota, E. (1987) *Proizvodnja i iskoriscavanje lekovitog bilja.* Nolit, Beograd, 139–143.

Konon, N.T. (1978) Biologia cvetenis i ophilenia valeriani lekharstvennoi v moskovskoi olasthi. *Rast.Resursi* **14**, 73–77.

Konon, N.T., Novikova, N.L. (1981) Reakcia valeriani lekh arstvennoi na inbriding. *Rast.Resursi,* **17**, 85–90.

Kuznechova, G.K., Konon, N.T., Sain, S.S., Pomanenko, V.I. (1984) Photostimulatia semian katharanthusa rozovogo i valeriani lekarsthvenoi predposevnim oblucheniem. *Konferencia Problemi photoenergetiki rastheni i povishenie urasainosti.* Tezis dokladov. (3–5 April, Livov) 160–161.

Lewkowicz-Mosiej, T. (1984) Problemy uprawy kozlka lekarskiego. *Wiad.Ziel.* **26**, 1–2.

Nedkov, N.K., Slavov, S.I. (1989) The effect of irrigation on valerian root yield. *Rasteniev'dni. Nauki.* **26**, 21–24.

Noller, P. (1989) *Untersuchung der Variabilitat von Valerensauren, Valerenal and Valeranon in Wildpopulationen und Zuchtmaterial von Valeriana officinalis* L. Justus-Liebig-Universitat, Gissen, Dissertation, p. 165.

Noller, P., Vömel, A. (1989) Breeding experiments as indications for the variance components of some cyclopentan-sesquiterpenes of *Valeriana officinalis* L. XII. *Eucarpia Congress, 1989. Vortrag für Pflanzenzüchtung,* 15.

Perrot, É., Paris, R. (1974) *Les Plantes Médicinales.* Presses Universitaires de France, 237–238.

Rácz, G., Rácz-Kotilla, E., Laza, A. (1984) *Gyógynövényismeret,* Ceres Könyvkiadó, Bukarest, 257–260.

Ruminska, A. (1973) *Rosliny lecznicze.* Panstwowe Wydawnictwo Naukowe, Warszawa, 293–309.

Sárkány, S., Baranyai, G. (1958) Die untersuchung der Arzneibaldriane in Ungarn. *Acta Bot. Acad.Sci.Hung.* **4**, 311–350.

Shugaeva, E. (1979) Male sterility in *Valeriana officinalis. Soviet.Genetics,* **15**, 93–97.

Sváb, J. (1978) *Macskagyökér. in Hornok L.: Gyógynövények termesztése és feldolgozása.* Mezõgazdasági Kiadó, Budapest, 152–157.

Titz, W., Jurenitsch, J., Gruber, J., Schabus, I., Titz, E., Kubelka, W. (1983) Valepotriate und atherisches öl morphologisch und chromosomal definierter Typen von *Valeriana officinalis* s.l. *Sci.Pharm.* **51**, 63–87.

Titz, W., Titz, E. (1982) Analyse der Formenmannigfaltigkeit der *Valeriana officinalis* - Gruppe in zentralen und südlichen Europa. *Ber. Deutsch.Bot. Ges.*, **95**, 155–164.

Vömel, A., Hölzl, J., (1979) Züchtungversuche durch Selektion und Isolation on *Valeriana officinalis* L. *Arzneipflanzen Colloquium in Rauischholzhausen*, 2.

Voroshilov, V.N. (1959) Lekharstvennaia valeriana. *Izd. Akad.Nauk,S.S.S.R.* Moskva, 60–139.

5. VALERIAN: QUALITY ASSURANCE OF THE CRUDE DRUG AND ITS PREPARATIONS

H.J. WOERDENBAG[1], R. BOS[1] AND J.J.C. SCHEFFER[2]

[1]*Department of Pharmaceutical Biology, University Centre for Pharmacy, Groningen Institute for Drug Studies (GIDS), University of Groningen, Antonius Deusinglaan 1, 9713 AV Groningen, The Netherlands*
[2]*Division of Pharmacognosy, Leiden/Amsterdam Center for Drug Research (LACDR), Leiden University, Gorlaeus Laboratories, P.O. Box 9502, 2300 RA Leiden, The Netherlands*

CONTENTS

ASSAY PROCEDURES

Extractives and Residues
Essential oil content
Tincture residue

Valepotriates

Spectrophotometry
Titrimetry
TLC
GC
HPLC

Valerenic Acid and Derivatives

TLC
HPLC

Baldrinals

TLC
HPLC

Essential Oil

GC
GC-MS

Mixed Constituents

TLC
HPLC

STORAGE CONDITIONS

REGULATORY ASPECTS

Present Situation
Future Developments

REFERENCES

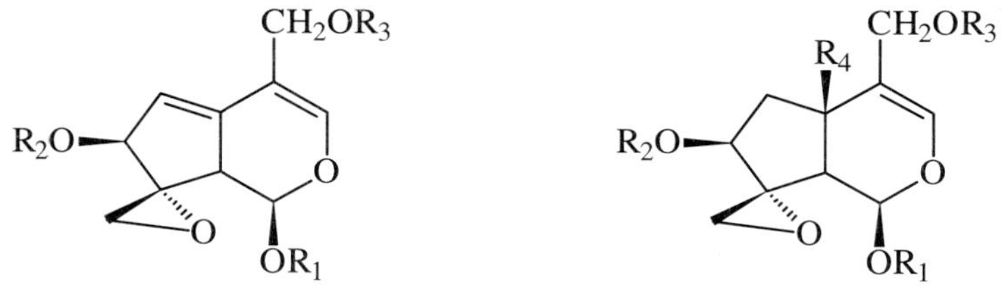

Valtrate
$R_1 = R_2 = COCH_2CH(CH_3)_2$
$R_3 = COCH_3$

Isovaltrate
$R_1 = R_3 = COCH_2CH(CH_3)_2$
$R_2 = COCH_3$

Acevaltrate
$R_1 = COCH_2C(CH_3)_2OCOCH_3$
$R_2 = COCH_2CH(CH_3)_2$
$R_3 = COCH_3$

Didrovaltrate
$R_1 = R_3 = COCH_2CH(CH_3)_2$
$R_2 = COCH_3$
$R_4 = H$

IVHD
$R_1 = COCH(OCOCH_2CH(CH_3)_2)CH(CH_3)_2$
$R_2 = COCH_2CH(CH_3)_2$
$R_3 = COCH_3$
$R_4 = OH$

Figure 1 Valepotriates found in *Valeriana* spp.

INTRODUCTION

The three most important *Valeriana* species used in herbal medicine in Europe are *Valeriana officinalis* L. *s.l.*, *V. wallichii* DC. (synonym *V. jatamansi* Jones; Indian or Pakistani valerian), and *V. edulis* Nutt. ssp. *procera* F.G. Meyer (synonym *V. mexicana* DC.; Mexican valerian). The roots and rhizomes of the three species are all used as mild sedatives, but show large differences with regard to their constituents. Consequently, phytomedicines prepared from these species are characterized by different chemical compositions (Hänsel 1990; Bos *et al.*, 1994, 1996a).

It is still not fully clear which constituents are to be held responsible for the sedative action, but the valepotriates (Figure 1) as well as valerenic acid and its derivatives (acetoxyvalerenic and hydroxyvalerenic acids) (Figure 2) are generally considered to contribute to it (Houghton 1988; Morazzoni and Bombardelli 1995; Bos *et al.*, 1996a). In addition, the essential oil may play a role in the biological activity of valerian (Hazelhoff *et al.*, 1979a, Bos *et al.*, 1994, 1996a). Based on their chemical structure, valepotriates are divided into two main groups: the diene type (valtrate, isovaltrate and acevaltrate) and the monoene type (didrovaltrate, isovaleroxyhydroxydidrovaltrate) (Steinegger and Hänsel 1992). Quality assurance and quality control of the crude drug and its preparations should therefore be based on these major groups of secondary metabolites.

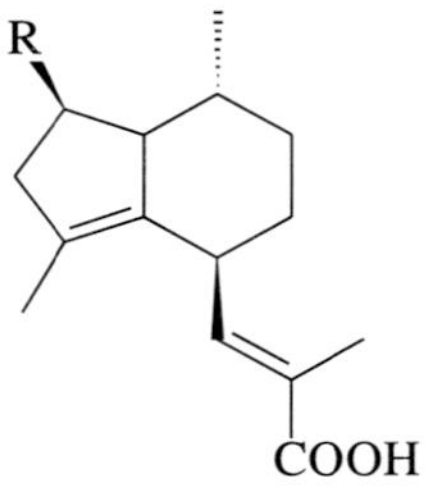

Valerenic acid
R = H

Hydroxyvalerenic acid
R = OH

Acetoxyvalerenic acid
R = OCOCH$_3$

Figure 2   Valerenic acid and related compounds found in *Valeriana officinalis*

V. officinalis contains valerenic acid and derivatives as well as valepotriates of the diene type, mainly valtrate and isovaltrate. Small amounts of didrovaltrate and isovaleroxyhydroxydidrovaltrate may also be present (Thies 1968; Titz *et al.,* 1982, 1983; Bos *et al.,* 1997a). *V. wallichii* and *V. edulis* lack valerenic acid and derivatives, but contain valepotriates, in higher contents than *V. officinalis*. In *V. wallichii*, valtrate, isovaltrate and didrovaltrate are found (Wienschierz 1978; Bos *et al.,* 1992). In *V. edulis*, valtrate, isovaltrate, acevaltrate, didrovaltrate and isovaleroxyhydroxydidrovaltrate are present (Thies 1968; Lorens 1989). Essential oil can be isolated from *V. officinalis* (Hazelhoff *et al.,* 1979a; Hendriks and Bos 1984; Bos *et al.,* 1986) and *V. wallichii* (Bos *et al.,* 1992). In *V. edulis* only trace amounts of volatiles are present (Hendriks and Bos 1984; Steinegger and Hänsel 1992; Bos *et al.,* 1997d). In Table 1 a survey is given of the contents of these groups of compounds in the three valerian species, as reported in the literature.

Table 1   Contents -based on dry weight- of the major groups of compounds in three valerian species used in herbal medicine (Bos *et al.,* 1997a)

Plant	valerenic acid (%m/m) and derivatives	valepotriates (%m/m)	essential oil (%v/m)
V. officinalis	0.1-0.5	0.8-1.7	0.5-2.0
V. wallichii	absent	1.8-3.5	0.1-2.0
V. edulis	absent	8.0-12.0	<0.1%[*]

[*]The volatile compounds that are obtained after distillation of the roots of this plant also include valeric, isovaleric and hydrovaleric acids, as well as several decomposition products formed upon heating of valepotriates

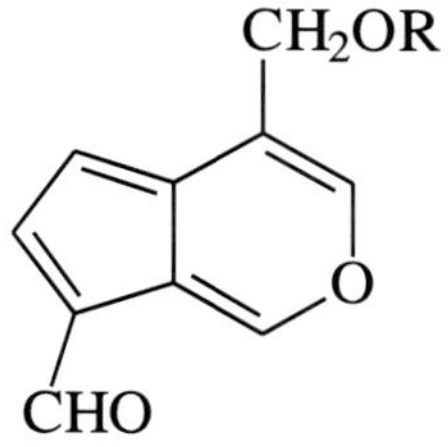

Baldrinal
R = COCH$_3$

Homobaldrinal
R = COCH$_2$CH(CH$_3$)$_2$

Figure 3 Baldrinal and homobaldrinal

All three species are used for the production of solid, oral dosage forms, while from *V. officinalis* also tinctures and a tea are made (Bos *et al.*, 1996a). Phytomedicines that are standardized on valepotriates are mostly prepared from *V. wallichii* and *V. edulis*, because these species are relatively rich in valepotriates (Wichtl 1989).

Another relevant group for analysis is formed by the baldrinals (Figure 3), yellow-coloured decomposition products of the valepotriates. Baldrinal originates from valtrate and acevaltrate; homobaldrinal from isovaltrate (Steinegger and Hänsel 1992; Bos *et al.*, 1997a). As cytotoxic and mutagenic effects have been described for the baldrinals (Bounthanth *et al.*, 1981; Braun *et al.*, 1986; Von der Hude *et al.*, 1986; Dieckmann 1988), their absence in the crude drug and in preparations has to be proved in order to avoid possible hazardous effects (Bos *et al.*, 1996a).

At present, there is some discussion about the safety of valepotriates as well. Due to the reactive epoxy group, valepotriates possess alkylating properties. Cytotoxicity and mutagenicity in *in vitro* cell systems have been described. The compounds have been shown to inhibit DNA and protein synthesis in *in vitro* cultured mammalian cells (Bounthanth *et al.*, 1981, 1983, Von der Hude *et al.*, 1985, 1986; Hänsel 1990, 1992; Keochanthala-Bounthanth *et al.*, 1990, 1993). It is yet unclear to what extent these toxic effects are relevant for humans after ingestion of valepotriate-containing preparations (Bos *et al.*, 1997a). However, more and more preference is given to valerian preparations that are devoid of the potentially hazardous valepotriates and baldrinals (De Smet and Vulto 1988; Hänsel 1992). Tinctures and teas, prepared from roots and rhizomes of *V. officinalis*, that is official in the European Pharmacopoeia 2nd edn, meet this demand (Hänsel and Schulz 1985: Schimmer and Röder 1992; Bos *et al.*, 1996a).

In this chapter analytical aspects related to the medicinally used valerian species are discussed. Pharmacopoeial aspects are dealt with, and analytical procedures to assay the different groups of secondary metabolites, as described in the literature, are presented and discussed.

PHARMACOPOEIAL QUALITY ASSURANCE - IDENTITY

Introduction

Crude drug

Pharmacopoeial aspects for the crude drug include a definition, and macroscopical and microscopical description of the crude drug, identity and purity reactions, and an assay for a quantitative determination.

Valerianae radix or valerian root is listed in the current European Pharmacopoeia, 2nd edn (1993). Hence, this monograph is also part of, among others, the German Pharmacopoeia, 'Deutsches Arzneibuch 10', the French Pharmacopoeia, 'Pharmacopée Française X', the British Pharmacopoeia 1993, and the Dutch Pharmacopoeia, 'Nederlandse Farmacopee IX'. *Valerianae radix* is defined as the subterranean organs of *Valeriana officinalis* L. *s.l.*, including the rhizome, roots and stolons, carefully dried at a temperature below 40°C. It contains not less than 0.5% (v/m) of essential oil (for assay see under 'Assay procedures (qualitative and quantitative determinations)').

Valerian was included in the second edition of the Indian Pharmacopoeia (1966) and defined as the dried rhizomes, stolons and roots of *Valeriana wallichii* DC. This monograph is not included in the current third edition. *V. edulis* has been included in the Mexican Pharmacopoeia. No monograph on valerian is found in the United States Pharmacopeia, USP 23 / National Formulary, NF 18 (1995).

Tinctures

The European Pharmacopoeia 2nd edn (1993) furthermore contains a general monograph on tinctures, that should be applicable to *Tinctura Valerianae* (valerian tincture). However, very limited information is included as yet. According to this pharmacopoeia, tinctures are prepared by maceration, percolation or other suitable, justified methods, using ethanol of a suitable concentration. Tinctures are obtained using 1 part of drug and 10 parts of extraction solvent or 1 part of drug and 5 parts of extraction solvent. Limits are given for methanol and 2-propanol: not more than 0.05% (v/v) of each of these alcohols is allowed. It is furthermore stated that a tincture should comply with limits prescribed for relative density, ethanol content and drug residue, however, without giving the limits.

According to the Dutch Pharmacopoeia 6th edn, 2nd printing (1966), valerian tincture is prepared by maceration from 1 part of valerian root and 5 parts of ethanol (70% (v/v)). The tincture has a brown colour. The relative density of the tincture is 0.897-0.907 (at 20°C). The drug residue is at least 2.5% (m/v).

According to the German Pharmacopoeia, 'Deutsches Arzneibuch', DAB 10 (1993), a valerian tincture is prepared by percolation of 1 part of valerian root with 5 parts of ethanol 70% (v/v). The ethanol content is 63.5-69.0% (v/v). The drug residue is at least 3.0% (m/v) (for assay see under 'Assay procedures (qualitative and quantitative determinations)').

Recently, we studied the influence of different ethanol/water ratios on the composition of valerian extracts made from *V. officinalis*. At an ethanol concentration of 30% (v/v), valerenic acid and its derivatives started to be extracted from the root material. Above 50% (v/v) ethanol the amounts of these sesquiterpenoids were more or less constant. Valepotriates were extracted only with ethanol concentrations above 70% (v/v). This means that if 70% (v/v) ethanol is used to prepare a valerian tincture, the extraction of the valepotriates is far from complete (Bos *et al.,* 1996a).

Botanical Aspects

Macroscopy

Valeriana officinalis

The European Pharmacopoeia 2nd edn (1993) describes the morphology of the crude drug as follows. The rhizome is yellowish-grey to pale greyish-brown, obconical to cylindrical, up to 50 mm long and up to 30 mm in diameter. The base is elongated or compressed, covered by and merging with numerous roots. The apex usually exhibits a cup-shaped scar from the aerial parts. Stem-bases are rarely present. In longitudinal section, the pith exhibits a central cavity traversed by septa. The roots are numerous, almost cylindrical, of the same colour as the rhizome, 1 mm to 3 mm in diameter and sometimes more than 100 mm long. A few filiform fragile secondary roots are present. The fracture is short. The stolons are pale yellowish-grey, showing prominent nodes separated by longitudinally striated internodes, each 20 mm to 50 mm long, with fibrous fracture.

Valerian root has a characteristic and penetrating odour, resembling that of valeric acid and camphor. The taste is somewhat sweet at first, then spicy and slightly bitter. Properly dried material has the odour of the fresh essential oil, without a note of valeric acid. Only poorly dried or old material is characterized by the valeric acid odour.

Valeriana wallichii

The Indian Pharmacopoeia 2nd edn (1966) describes the macroscopical characters of the crude drug as follows. Indian valerian consists of dull yellowish-brown rhizomes, 4-8 cm long and 4 to 10 mm thick, and a very variable amount of roots up to 7 cm long and 1-2 mm thick. The rhizomes are unbranched and somewhat flattened dorsiventrally. The upper surface bears leaf scars and the lower surface roots or root scars. The rhizome breaks with a short fracture, and the horny interior shows a small dark bark, a well-marked cambium, about 12-15 light-coloured xylem bundles and a dark pith and medullary rays. Stolons connect the rhizomes, are stout, 1 to 5 mm long and 2 to 4 mm thick, yellowish grey in colour, longitudinally wrinkled usually with nodes and internodes and bearing adventitious roots. Occasionally thin stolons, 1 to 2 mm thick, are found. Roots are yellowish-brown, 3 to 5 cm long and 1 mm thick. The odour is strong and reminiscent of isovaleric acid. The taste is bitter and somewhat camphoraceous (Evans 1989).

Valeriana edulis

No macroscopical description is known.

Microscopy

Valeriana officinalis

The European Pharmacopoeia 2nd edn (1993) describes the microscopical characters of the crude drug as follows. The transverse section of the root shows small, suberised epidermal cells, some with root hair. The exodermis consists of one or occasionally two layers of larger, suberised cells, often containing droplets of essential oil. The outer cortex comprises two to four layers of resin-containing cells with thin collenchymatous, sometimes suberised walls. The inner cortex is composed of numerous layers of polygonal to rounded cells filled with starch. The starch granules are simple or compound. The simple granules are rounded, 5-15 μm in diameter, sometimes showing a cleft or stellate hilum. The compound granules, with two to six components, are up to 20 μm in diameter. The endodermis consists of a single layer of suberised, tangentially elongated cells. The pericycle is continuous and starch-filled. Parenchyma surrounds the phloem zone. The cambium is frequently indistinct. The vascular bundles form an interrupted ring surrounding the starch-filled cells.

The rhizome in transverse section has a different anatomy from the root. Its structure is complicated by the presence of numerous vascular bundles coming from the roots and stolons. The epidermis and exodermis are partly replaced by poorly developed periderm. The central pith is wide and has cavities of various sizes, the larger ones being separated by plates of partially sclerified tissue.

The powder is light brown and is characterized by numerous fragments of parenchyma with rounded or elongated cells and containing starch granules as described above. Also present are cells containing light-brown resin; rectangular sclereids with pitted walls, 5-15 μm thick; xylem, isolated or in non-compact bundles, 10-50 μm in diameter; some absorbing root hairs and cork fragments.

It is difficult to see the difference under the microscope between the pulverized roots of the three valerian species. Thin-layer chromatography, to determine the presence of valerenic acid, is the best method to prove the identity of *V. officinalis* (Wichtl 1989).

Valeriana wallichii

The Indian Pharmacopoeia 2nd edn (1966) describes the microscopical characters of the crude drug as follows. The rhizome-cork consists of 4 to 14 layers of lignified suberised cells, that occasionally contain oil globules. The cortex is parenchymatous and contains numerous starch grains, oil globules an a yellowish-brown substance. The outer 2 or 3 layers of cortex are collenchymatous and occasional root traces appear as paler strands. The endodermis is one layered. The pericycle is parenchymatous and within it 12 to 18 collateral vascular bundles, separated by dark medullary rays, are present. The pith is large, parenchymatous, lacunar and contains starch grains. Starch occurs as single or occasional compound grains of two components, individual grains being 7 to 30, mostly 10 to 25 μm, in diameter. Calcium oxalate is absent.

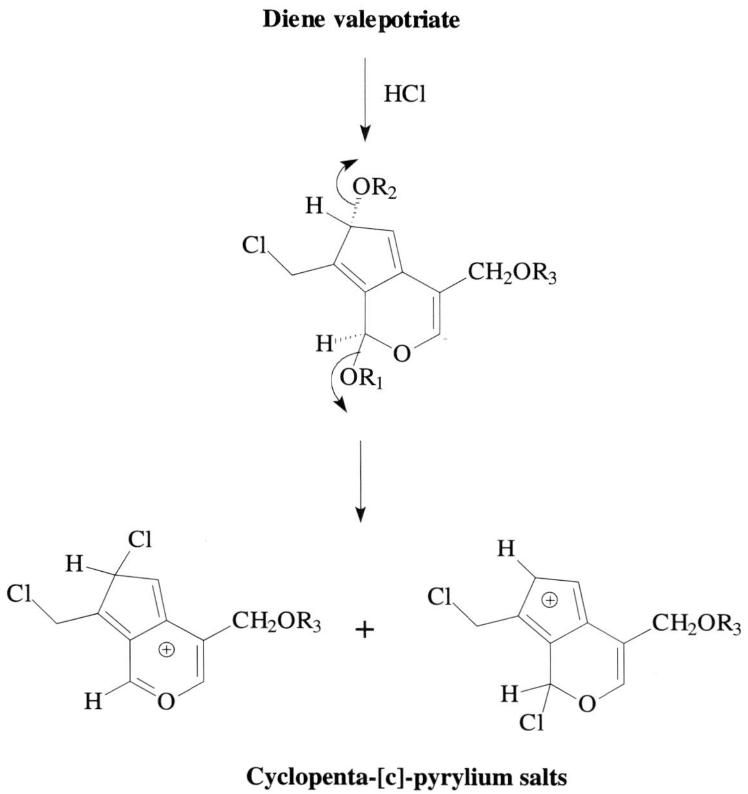

Diene valepotriate

Cyclopenta-[c]-pyrylium salts

Figure 4 Pseudoazulene formation from valepotriates acid conditions

Stolon-cork consists of 2 to 5 layers; the cortex up to 25 layers, parenchymatous, followed by about 20 collateral vascular bundles, which in young stolons are separated by cellulosic parenchymatous medullary rays, which in older stolons become lignified. Pith wide and lacunar. Root traces are absent. Roots have small central parenchymatous pith surrounded by tetrarch to polyarch xylem and a wide parenchymatous bark.

Valeriana edulis

No microscopical description is known.

Chemical Tests

Crude drug

The identity test for *V. officinalis* in the European Pharmacopoeia 2nd edn (1993) consists of the addition of a mixture of equal volumes of glacial acetic acid (98% (m/m)) and hydrochloric acid (25% (m/v)) to a methylene chloride extract of the freshly powdered drug and the formation of a blue colour within 15 minutes.

This test shows the presence of valepotriates. Valepotriates are lipophilic compounds that are extracted from the drug by methylene chloride. With the acetic acid-hydrochloric acid mixture blue coloured cyclopenta-[c]-pyrylium salts (pseudoazulenes) are formed from valepotriates with a conjugated diene structure (valtrate, isovaltrate, acevaltrate) (Böhme and Hartke 1979). The proposed structure of these salts is shown in Figure 4 (Thies 1968, 1969).

This identity reaction, however, is not specific for *V. officinalis*. The roots of other *Valeriana* species, which also contain valepotriates of the diene type, will give a positive reaction as well (Hartke and Mutschler 1987). A conclusion about the identity, however, will be possible after other tests have been carried out. The identity of *V. officinalis* can be based on the purity test of the European Pharmacopoeia 2nd edn (1993), using TLC (see below). In that test the presence of valerenic acid and derivatives, which are characteristic of *V. officinalis*, is checked. In addition, HPLC methods are available to determine the characteristic valepotriate composition of the crude drug, as well as valerenic acid and its derivatives. These assays, however, are not (yet) included in the European Pharmacopoeia.

Tinctures

The German Pharmacopoeia, DAB 10 (1993), determines the identity and purity of valerian tincture by TLC using silica gel and glacial acetic acid (98% (v/v))-ethyl acetate-hexane 0.5:35:65 as the mobile phase. The tincture is treated with aqueous KOH and washed with methylene chloride. The remaining aqueous layer is acidified and extracted with methylene chloride to provide the test solution for chromatography. Methanolic solutions of fluorescein and Sudan red G are used as references. Detection after development is effected by spraying with anisaldehyde/sulphuric acid reagent.

The chromatogram of the test solution shows an intense blue zone at about the R_f of fluorescein due to hydroxyvalerenic acid and a violet zone at about the R_f of Sudan red G of valerenic acid. In the upper part of the chromatogram several other faint red to violet coloured zones are found.

In a tincture prepared from *Valeriana* species other than *V. officinalis*, the zones of valerenic acid and hydroxyvalerenic acid are absent.

This TLC method is similar to that described by Hänsel *et al.* (1983), with the difference that these authors used a solution of 1 mg vanillic acid and 2 mg anisaldehyde in methanol as the reference solution. To valerenic acid an R_f value of 0.4-0.5 (just above anisaldehyde) was assigned, and to hydroxyvalerenic acid an R_f value of 0.07-0.12 (just below vanillic acid).

Valerenic acid and acetoxyvalerenic acid, that are characteristic of *V. officinalis*, can be detected immediately in the tincture. However, a purification step is added to the procedure. Tinctures that are prepared from Indian or Mexican valerian contain neutral substances that can disturb the detection of the two acids. The acids are converted into their corresponding potassium salts. The disturbing neutral compounds can now be removed from the aqueous phase by extraction with methylene chloride. Acetoxyvalerenic acid is hydrolysed in the aqueous phase by warming, yielding hydroxyvalerenic acid.

After acidification, the sesquiterpene carboxylic acids are extracted with methylene chloride. This procedure to check the identity of valerian tincture can also be used to prove its purity, because falsifications are recognized with this procedure. It should be noticed that hydroxyvaleric acid is not a genuine constituent of *V. officinalis*; the acid originates from acetoxyvaleric acid after alkaline hydrolysis.

The German Homoeopathic Pharmacopoeia, 'Homöopatisches Arzneibuch', HAB 1 (1975) identity test comprises a TLC identity reaction for a stored mother tincture, that differs from a freshly prepared mother tincture (see above). Valepotriates are not detectable in a stored tincture. Here, constituents of the essential oil are analysed.

The system used consists of silica gel with methylene chloride as the mobile phase. The tincture is applied without further modification and, as a reference solution, 10 mg of borneol and 10 mg of bornyl acetate are dissolved in 10 ml of methanol. Detection after development is effected by spraying with anisaldehyde/sulphuric acid reagent. The chromatogram of the reference solution shows a brown-violet zone of borneol in the lower part and a brown-violet zone of bornyl acetate in the upper part. Related to bornyl acetate (RR_f 1.0), borneol has an RR_f of 0.4. In the chromatogram of the test solution violet zones with RR_f values of 0.3, 0.8 and 1.1 as well as a red zone with an RR_f of 1.6, related to borneol (RR_f 1.0) and violet zones with RR_f values of 0.8, 1.0, 1.2 and 1.5, related to bornyl acetate (RR_f 1.0) are found.

PHARMACOPOEIAL QUALITY ASSURANCE - PURITY

Chromatographic Tests

Thin-layer chromatography (TLC)

According to the European Pharmacopoeia 2nd edn (1993) the purity of the crude drug *Valerianae radix* is determined by TLC, using silica gel and a mixture of ethyl acetate-hexane 3:7 as the mobile phase.

As a test solution the same methylene chloride extract as used for the identity reaction (see above) is employed and methanolic solutions of aminoazobenzene and Sudan red G are used as references. Double development is employed and detection is achieved after spraying with anisaldehyde/sulphuric acid reagent.

The chromatogram obtained with the test solution should show in the middle, at an R_f value between those of the pink zone (Sudan red G) and of the orange zone (aminoazobenzene) in the chromatogram of the reference solution, a deep-violet zone (valerenic acid) and sometimes above this zone a greyish-brown zone (valtrate and isovaltrate). Furthermore, a faint violet zone (acetoxyvalerenic acid) with an R_f value lower than that of the zone due to aminoazobenzene is seen, as well as grey zones situated between the zone due to valerenic acid and the starting point, a number of violet zones of variable intensity in the upper part of the chromatogram, and a mostly very faint violet zone immediately above the starting point.

Because valepotriates as well as valerenic acid and derivates were not commercially available as reference substances until recently, the location of the spots due to these compounds are related to the location of Sudan red G and aminoazobenzene. Nowadays (iso)valtrate and valerenic acid can be purchased. The purity test is selective for *V. officinalis*, because the presence of valerenic acid and derivatives is checked (Hartke and Mutschler 1987). In addition, essential oil components are also separated using this system.

The chromatographic tests prescribed to establish identity also serve as a method whereby impurities, particularly species related to the required *Valeriana* sp., can be detected. In the German Homoeopathic Pharmacopoeia, 'Homöopatisches Arzneibuch', HAB 1 (1975), a TLC method is described as a purity test for *Valerianae radix* as well as for a freshly prepared mother tincture. The same test was described in the European Pharmacopoeia 1st edn (1975). Here, silica gel with a fluorescence indicator (GF_{254}) is used as the stationary phase with hexane-methyl ethylketone 8:2 as the mobile phase which is double-developed. A methylene chloride extract of the drug is used, or the mother tincture, and solutions of vanillin and anisaldehyde are used as references.

When examined under light at 254 nm, the chromatogram obtained for the test solution displays a number of dark areas against a light background. The largest of these, corresponding to valtrate, has about the same R_f (0.5-0.6) as the area of anisaldehyde obtained using the reference solution. Subsequently, the plate is sprayed with dinitrophenylhydrazine-aceto-hydrochloric solution and heated at 100-105°C for 10 min. In the chromatogram obtained for the test solution a number of coloured areas appear. The area corresponding to valtrate is now coloured greenish-grey and that corresponding to anisaldehyde is yellow. The chromatogram obtained for the test solution shows, in the lower part, a blue area having the same R_f of about 0.3 as the yellow area corresponding to vanillin in the chromatogram obtained with the reference solution. The chromatogram obtained for the test solution shows, between the blue area and the area corresponding to valtrate, two much smaller and more weakly coloured areas (didrovaltrate and acevaltrate).

The above described analysis for purity can be used to investigate a possible contamination of *V. officinalis* with the other species; this might be done deliberately in order to enhance the valepotriate content. *V. officinalis* mainly contains valepotriates of the diene type (valtrate, isovaltrate and some acevaltrate), whereas both *V. wallichii* and *V. edulis* also contain considerable amounts of monoene valepotriates (didrovaltrate and isovaleroxyhydroxydidrovaltrate). Valepotriates with a diene structure appear as blue to greenish-grey colours; monoene valepotriates as yellow to greenish-grey colours (Böhme and Hartke 1979).

The blue area in the chromatogram obtained for the test solution, having the same R_f of about 0.3 as the yellow area corresponding to vanillin in the chromatogram obtained for the reference solution, is to be ascribed to isovaleroxyhydroxydidrovaltrate, according to Stahl (1970).

V. wallichii and *V. edulis* may contain valepotriate hydrins, yielding violet-blue zones in the chromatogram that are located immediately under the blue zone mentioned in the previous paragraph (Böhme and Hartke 1979). Valepotriate hydrins have the same chemical structure as valepotriates, but an acid is added to the epoxy group (Thies 1968).

Van Meer *et al.* (1977) pointed to the fact that, next to the valepotriates located in the lower half of the chromatogram, components of the essential oil of valerian root are present as well, which are located in the upper half, i.e. above the zone due to valtrate. The essential oil components can be distinguished from the valepotriates by first spraying the chromatogram with a mixture of glacial acetic acid-hydrochloric acid (25%) 1:1. The essential oil components colour purple to violet, whereas valepotriates yield green-grey (valtrate, acevaltrate), blue-green (isovaleroxyhydroxydidrovaltrate) and blue-grey spots (unidentified valepotriates with lower R_f values). After subsequent spraying of the chromatogram with dinitrophenylhydrazine reagent, two components of the oil (aldehydes and/or ketones; R_f 0.65 and 0.80) are coloured yellow to yellow-orange. After a while, two other components (R_f 0.88 and 0.95) that do not belong to the essential oil colour yellow. These components were not seen before, in daylight, under UV light or after spraying with acid. After spraying with the acid, under valtrate also purple zones are visible.

Other Tests

Extractives

According to the European Pharmacopoeia 2nd edn (1993) the extractable matter yield using 96% ethanol for *V. officinalis* should be not less than 15.0% w/w. The Indian Pharmacopoeia 2nd edn (1966) uses 60% alcohol for *V. wallichii* and specifies a minimum value of 30% m/m.

In the European Pharmacopoeia 1st edn (1975), this demand was included in the definition of valerian root. However, the determination of the amount of extractable matter is not a substitute for an assay, quantitating valepotriate and/or valerenic acid and its derivatives (Böhme and Hartke 1978). Moreover, after harvesting the fresh roots are washed. If the washing procedure has taken too long, the roots can leach, resulting in a lower amount of extractable matter (Hartke and Mutschler 1987).

Ash values

According to the European Pharmacopoeia 2nd edn (1993) the sulphated ash should not be more than 15.0% m/m.

When organic matter is incinerated as such, the residue found will depend of the temperature used. For instance, alkali chlorides and earth-alkali carbonates are volatile at certain temperatures. In the presence of sulphuric acid non-volatile sulphates are formed. Because pyrosulphates may be formed during the heating procedures, ammonium carbonate is added at the end. The sulphated ash is a measure for the total amount of inorganic matter in the plant material (Böhme and Hartke 1978).

Ash insoluble in hydrochloric acid

The maximum level set by the European Pharmacopoeia 2nd edn (1993) is 7.0%.

The ash insoluble in hydrochloric acid is the residue obtained after extracting the sulphated ash with hydrochloric acid, calculated with reference to 100 g of drug. Non-volatile, inorganic impurities are determined, such as soil and sand. It is a control of the washing procedure of the crude drug (Hartke and Mutschler 1987).

ASSAY PROCEDURES

Extractives and Residues

Essential oil content

The European Pharmacopoeia 2nd edn (1993) requires a minimum of 0.5%v/m for volatile oil in V. officinalis root. This test is a hydrodistillation in a special apparatus where the distillate is collected in a graduated tube.

Tincture residue

The determination of the residue from evaporation of the tincture is described in the European Pharmacopoeia 2nd edn (1993).

Results of drug residue determinations, obtained in our laboratory for the three *Valeriana* species discussed in this chapter: *V. officinalis* 28-31%, *V. wallichii* 21-46%, *V. edulis* 29-30%.

Valepotriates

Spectrophotometry

The first procedure for a quantitative determination of valepotriates has been described by Mannestätter *et al.* (1968). They isolated 'Halazuchrom B' (= valtrate) from a valerian extract, prepared with diethyl ether, after TLC on silica gel with methylene chloride-methyl ethyl ketone 9:1, or after column chromatography separation with hexane. Valtrate was measured spectrophotometrically at 254 nm.

Valtrate was also converted into a deep-blue coloured product using hydrochloric acid in methanol ('Halazuchrom-Reaktion') whereafter the absorbance was measured at 610 nm (Mannestätter *et al.,* 1968). The identity reaction for *V. officinalis*, as described in the European Pharmacopoeia 2nd edn (1993), is derived from this work.

As only diene valepotriates colour blue with hydrochloric acid, didrovaltrate cannot be determined in this way. In addition, didrovaltrate does not have a UV maximum at 254 nm. Wagner *et al.* (1970) described a spectrophotometric method by which all valepotriates can be determined. After a TLC separation of the valepotriates, followed by extraction, the hydroxylamine-iron(III) chloride reaction was carried out.

Because of their epoxide structure valepotriates are able to alkylate the nucleophilic agent 4-(4')-nitrobenzylpyridine (NBP). This reaction gives a linear response with time and dose and can therefore be used for a specific quantitative evaluation of valepotriates using a suitable standard (Braun *et al.,* 1983). After reaction of samples with NBP a solution of tetrene (tetramethylene pentamine) in acetone (2+3) is added and, after mixing, the absorbance was measured after 30 seconds at 560 nm.

Titrimetry

The first titrimetric determination of valepotriates has been described by Mannestätter *et al.* (1968), as the 'Jodhydrin-Methode'. The epoxy group of valepotriates was opened with sodium iodide in an ethanolic, acetate-buffered, acetic acid solution and the corresponding iodine hydrine was formed.

As this reaction proceeds very slowly (24 h), Lipták and Verzár-Petri (1980) published an improved titrimetric method with respect to time. The titration of valepotriates was performed by opening the epoxide ring with hydrobromic acid using a methylene chloride extract of the crude drug. This was put in the dark for 1.5 h. Subsequently, the flask was put on ice and the reaction mixture was rapidly titrated with 0.01 N sodium acetate solution. Just before the end of the titration 5 ml of a crystalviolet solution was added as an indicator. The titration was continued until a blue colour appeared.

Thin-layer chromatography (TLC)

Several TLC methods for the determination of valepotriates, qualitative as well as quantitative, are found in the literature. Stahl and Schild (1969) published the first complete working instruction for quickly testing valerian root for its valepotriates. They used a methylene chloride extract of plant material and silica gel with a fluorescence indicator (GF_{254}) with double development using hexane-methyl ethyl ketone 8:2 as the mobile phase. The spots were visualized under UV light (254 nm) and after spraying with benzidine-hydrochloric acid reagent (0.1% benzidine in hydrochloric acid (25%)-glacial acetic acid 1:1), followed by 160 min heating at 105°C.

Under UV light, the largest zone is found at R_f 0.5-0.6 (equal to anisaldehyde), due to valtrate. After the reaction with benzidine a series of coloured zones are seen. Valtrate becomes green-grey and anisaldehyde yellow. In the lower part of the chromatogram a blue zone is found at R_f 0.2-0.3 (equal to vanillin, yellow). Between the blue zone and the valtrate zone, two smaller and less intensively coloured zones are visible, due to didrovaltrate and acevaltrate.

The applied spray reagent was regarded as an improvement as compared with older reagents. Spraying with 6 N hydrochloric acid (Mannestätter *et al.,* 1967) yielded an unstable blue colour with valtrate and acevaltrate. Also a mixture of equal volumes of hydrochloric acid (25%) and glacial acetic acid only coloured the diene valepotriates (Thies and Funke 1966). Antimony trichloride 22% in chloroform (Thies and Funke 1966; Mannestätter *et al.,* 1967) did not give clear colours in daylight.

Benzidine is a carcinogenic agent; as a substitute, Stahl and Schild (1969) proposed a solution of 0.1% 2,4-dinitrophenylhydrazine in hydrochloric acid (25%)-glacial acetic acid 1:1 as a spraying reagent. As a disadvantage, a yellow background was obtained and the fluorescence was less. As an advantage, the intensity of the colours due to valtrate, acevaltrate (blue) and didrovaltrate (faint orange) was stronger.

Other mobile phases reported to give a good separation of valepotriates include benzene-ethyl acetate 9:1, light petrol-ethyl acetate 8:2, diethyl ether-hexane 5:5, light petrol-acetone 8:2 (Stahl and Schild, 1969).

Laufer *et al.* (1970) developed an improved TLC method for valepotriates, based on the method of Stahl and Schild (1969). As a stationary phase silica gel with a fluorescence indicator (GF_{254}) was used. *n*-Hexane-methyl ethyl ketone 8:2 was used as the mobile phase. The reference solution was the same as applied by Stahl and Schild (1969), or was composed of reference valepotriates, 20 mg of valtrate, 20 mg of acevaltrate and 20 mg of didrovaltrate in 10 ml methanol. The plates were developed twice in a saturated chamber over a path of 15 cm. Under UV light of 254 nm valtrate and acevaltrate are visible as purple spots against a greenish fluorescent background. Spraying was done with a 22% antimony chloride solution, immediately followed by the 2,4-dinitrophenylhydrazine reagent. After 5 min at 105°C, valtrate, acevaltrate and isovaleroxyhydroxydidrovaltrate turn blue and didrovaltrate orange. The colour of the spots as well as the reagent are stable. Quantitative valtrate determinations were done directly on the thin-layer plate by using a densitometer (at 258.5 nm).

Verzár-Petri et al. (1976) applied a two-dimensional TLC technique for the separation of valepotriates using hexane-methyl ethyl ketone 8:2, hexane-methyl ethyl ketone 9:1 and methylene chloride-methyl ethyl ketone 9:1 as mobile phases. Detection was done under UV light of 254 nm and after spraying with dinitrophenylhydrazine reagent.

Later, a better separation of the valepotriates was achieved by Hazelhoff et al. (1979b), using toluene-ethyl acetate-methyl ethyl ketone 80:15:5 as the mobile phase. This eluent was used for qualitative and quantitative TLC, in an unsaturated chamber, run 13 cm. Detection was achieved under UV light of 254 nm and with dinitrophenylhydrazine reagent.

Rücker et al. (1981) described a TLC method by which valepotriates can be determined next to baldrinals. HPTLC RP-18 F_{254} and HPLTC-silica gel 60 F_{254} plates (Merck, Darmstadt, Germany) were used. Mobile phase was methanol-water 8:2 for the reversed-phase plates, and methylene chloride-ethyl acetate-acetone 48:1:1 or methylene chloride-methyl ethyl ketone 98:2 for the straight-phase plates. Ammonia vapour was used as detection reagent. The reaction products of ammonia with valtrate, isovaltrate, acevaltrate, baldrinal and homobaldrinal could also be used for quantitative densitometric determinations. Valtrate and isovaltrate could be assayed without a detection reagent at 255 nm, the baldrinals without a detection reagent at 425 nm, and didrovaltrate after reaction with ammonia at 279 nm. Following exposure to ammonia vapour, valtrate and didrovaltrate could be assayed again, at 279 nm. This method was about ten-fold more sensitive than the detection with anisaldehyde reagent (Stahl and Schütz, 1980).

The reaction with NBP has been used by Braun et al. (1983) for the quantitative determination of valepotriates (see under 'Spectrophotometry'). In addition, the NBP reaction was shown to be useful for the detection of valepotriates by TLC. After separation of valerian root extracts on silica gel with a fluorescence indicator using n-hexane-methyl ethyl ketone 8:2 as the mobile phase, the plates were sprayed with a 3% solution of NBP in acetone. After heating the plate at 40°C for 90 min, the plate was submerged in a 10% solution of tetrene (tetramethylene pentamine) in acetone. Valepotriates yield deep-blue to violet-blue spots.

Gas chromatography (GC)

Verzár-Petri et al. (1976) described the first GC method for the determination of valtrate, acevaltrate and didrovaltrate. In addition, baldrinal could be detected. They used two different stationary phases: 2% OV-1 and 3% OV-17.

A GC determination of valepotriates has also been described by Graf and Bornkessel (1978). The substances, extracted by methylene chloride, were first separated using TLC. The valepotriates were then eluted from their respective zones, and hydrolysed with a

0.5 N methanolic potassium hydroxyde solution in methanol (60 min at 70°C). Each valepotriate yielded isovaleric acid after the alkaline hydrolysis, that was subsequently extracted with diethyl ether containing propionic acid as the internal standard. Isovaleric acid was determined by GC on a Carbowax column.

High-performance liquid chromatography (HPLC)

The first HPLC method for the determination of valepotriates has been published by Tittel and Wagner (1978) and Tittel *et al.* (1978). They obtained a good separation suitable for qualitative as well as quantitative purposes from both crude material and preparations. A silica gel column with a particle size of 10 µm (MN-Nucleosil 50, 25 cm x 4 mm i.d., or Lichrosorb Si 100, 35 cm x 8 mm) and *n*-hexane-ethyl acetate 20:3 as the eluent were used. For detection a refractive index detection system and UV detection were applied. The authors stated that this method was superior to the combined TLC-spectrophotometric method using the hydroxylamine-iron(III)chloride reagent (Wagner *et al.,* 1970), with respect to time consumption, precision and sensitivity. The sequence of elution was valtrate and isovaltrate, didrovaltrate, acevaltrate.

For the determination of (iso)valtrate another HPLC method has been presented by Hazelhoff *et al.,* (1979b). A Spherisorb Silica S5W column, 250 mm x 4.6 mm i.d., particle size 5 µm (Chrompack, Middelburg, The Netherlands), with 0.8% methanol in hexane as the mobile phase, and detection at 254 nm were applied.

Van Meer and Labadie (1981) described a straight-phase as well as a reversed-phase HPLC method for the analysis of valepotriates. The straight- phase column was a 5 µm Partisil silica column (Chrompack), 25 cm x 3 mm i.d. As eluents *n*-hexane-ethyl acetate 90:10 or 95:5, or *n*-hexane-ethanol 99.5:0.5 were used. The reversed-phase column was a 5 µm Spherisorb ODS (Chrompack), 25 cm x 3 mm i.d. Methanol-water 50:50 or acetonitrile-water 60:40 were the eluents applied. Monoene valepotriates were detected at 206 nm, diene valepotriates at 256 nm.

Dossaji and Becker (1981) used reversed-phase HPLC with gradient elution for the determination of the e valepotriates. A Waters µ Bondapak C_{18} column, 30 cm x 3.9 mm i.d., was used with a methanol-water mixture as the eluent: A: 20:80 and B: 80:20; initially 85% B, finally 100% B in 5 min. Monoene valepotriates were separated on the same column, isocratic with methanol-water 80:20. Detection was done at 254 nm (dienes) and 208 nm (monoenes).

A comparable system was presented by Förster *et al.* (1984). The authors used an Altex Ultrasphere-ODS preparative column, 25 cm x 10 mm, particle size 5 µm, equipped with an Altex guard column (45 x 4.6 mm, 30 µm particles pellicular ODS). Monoenes were eluted with acetonitrile-water 70:30 and detected at 206 nm. For dienes a gradient was used of methanol-water mixtures, A: 60:40 and B: 90:10; initially 60% B, finally 90% B in 15 min. Detection of the dienes at 254 nm.

Chavadej *et al.* (1985) applied gradient elution for the monoenes as well as the dienes. A Lichrospher 100 CH (5 µm) column (Merck) was used. A 10 min gradient elution was achieved with methanol-water, A: 40:60 and B: 90:10; starting with 75% B to 95% B. The dienes were detected at 254 nm and the monoenes at 206 nm. As an internal standard *n*-pentylbenzene was used.

Valtrate and isovaltrate have been analysed in a methylene chloride extract of valerian, on a Lichrosorb RP 18 column (7 μm) using methanol-water 80:20 with 0.5% phosphoric acid as the eluent. Detection was done at 255 nm. With a flow rate of 2 ml/min, isovaltrate eluted after 6.4 min and valtrate after 7.4 min (Hänsel and Schulz 1985).

Gränicher *et al.* (1992) applied the system of Förster *et al.* (1984), but with some modifications. They used a Nucleosil C_{18} column (25 cm x 4 mm i.d., 5 μm) fitted with a Nucleosil C_{18} guard-column (30 x 4 mm i.d., 5 μm). Pentylbenzene was used as internal standard. The solvent system was methanol-water 90:10 (pump A) and methanol-water 40:60 (pump B). Elution of the diene type valepotriates was achieved isocratic with 60% A for 15 min, and a linear gradient to 90% A in 22 min. The monoene valepotriates were eluted with 60% A for 27 min, followed by a linear gradient to 80% A in 14 min.

Valerenic Acid and Derivatives

Initially, analytical procedures were focussed on the valepotriates as they were considered to be the main pharmacologically active constituents of valerian. In the 1980s, however, the valerenic acids started to attract attention and analytical procedures for these compounds, only present in *V. officinalis*, were developed.

Thin-layer chromatography (TLC)

For a better detection of acetoxyvalerenic acid in valerian root, Schimmer and Röder (1992) proposed to extend the TLC purity test described in the German Pharmacopoeia, DAB 9, (also the procedure described in the European Pharmacopoiea 2nd edn (1993)) with the TLC identity test for valerian tincture as included in the same pharmacopoiea since acetoxyvalerenic acid could better be determined after conversion into hydroxyvalerenic acid.

In addition, Schimmer and Röder (1992) obtained a better separation of the zone due to valerenic acid from the tincture following a second development of the plate, using toluene-ethyl acetate 93:7.

High-performance liquid chromatography (HPLC)

Hänsel and Schulz (1982) were the first to describe an HPLC method for the determination of valerenic, acetoxyvalerenic and hydroxyvalerenic acids, and valerenal, in both the crude drug and tinctures of *V. officinalis*. The co-occurrence of the two sesquiterpenoids valerenic and acetoxyvalerenic acid, and the possible presence of hydroxyvalerenic acid, is confined to *V. officinalis* and this HPLC method can therefore be employed to detect adulterations by non-officinal valerian root extractives (*V. wallichii* and/or *V. edulis*).

The HPLC method consisted of a reversed-phase analysis using an RP 18 (7 μm) column (Knauer, Berlin, Germany), 25 cm x 4.6 mm i.d., and a guard column of 40 mm. A mixture of methanol-water 80:20 with 0.5% phosphoric acid was used as the eluent (pH 2). Detection was done at 225 nm. It was shown that valepotriates had retention times that were comparable with those of the sesquiterpenoids, disturbing the analysis of root material. The lipophilic acids were separated from the lipophilic neutral components by treatment with alkali. After acidification of the alkaline aqueous fraction containing the sesquiterpenoids, they were extracted with chloroform. Tinctures of *V. officinalis* were analysed directly, as they were devoid of valepotriates. Biphenyl was used as an internal standard.

Freytag (1983) presented a simplified and more reproducible procedure when compared with the method of Hänsel and Schulz (1982). Of crude root material (*V. officinalis*), 2 g were extracted with methylene chloride using a Soxhlet apparatus. The solvent was evaporated, and the residue dissolved in methanol and transferred to a 100 ml volumetric flask. This extract contained both valerenic acids and valepotriates. The same procedure, but with ethanol-water 70:30 with 1% acetic acid yielded an extract free of valepotriates, whereas valerenic acids were quantitatively extracted. HPLC was done using a Lichrosorb RP 18 (5 μm) column, 250 x 4.6 mm i.d., and acetonitrile-phosphoric acid (pH 2.0) 65:35 as the mobile phase. The compounds were detected at 225 nm. Using this procedure, acetoxyvalerenic acid was determined as such, instead of in the form of hydroxyvalerenic acid. No alkaline-acid liquid-liquid extraction was necessary to separate valerenic acids from the valepotriates.

Baldrinals

Thin-layer chromatography (TLC)

A TLC method for the determination of baldrinal and homobaldrinal (next to valepotriates) is described by Rücker *et al.* (1981). HPTLC RP-18 F$_{254}$ and HPLTC silca gel 60 F$_{254}$ plates (Merck) were used. Mobile phase was methanol-water 8:2 for the reversed-phase plates and methylene chloride-ethyl acetate-acetone 48:1:1 or methylene chloride-methyl ethyl ketone 98:2 for the straight-phase plates. Ammonia vapour was used as detection reagent.

High-performance liquid chromatography (HPLC)

Baldrinals can be assayed using the HPLC method of Bos *et al.* (1996a), simultaneously with valerenic acid and derivatives and with valepotriates (see below).

Essential Oil

After isolation by hydrodistillation, the composition of the essential oils of *V. officinalis*, *V. wallichii* and *V. edulis*, respectively, has been studied by gas chromatography (GC) and gas chromatography coupled with mass spectrometry (GC-MS) (Bos *et al.,* 1996b, 1997b, c, d).

The oils are isolated from plant material by steam distillation and diluted 50 times with cyclohexane prior to GC analysis. In addition, each oil was separated into two fractions, with hydrocarbons and oxygen containing compounds, respectively, by eluting 250 μl of oil on a silica gel solid phase extraction column with subsequently 5 ml *n*-hexane and 5 ml diethyl ether. After gentle evaporation of the solvents of both fractions, 50 μl of each residue was diluted with 950 μl cyclohexane, and submitted to GC and GC-MS analysis.

Table 2 Main constituents of the essential oils of *V. officinalis*, *V. wallichii* and *V. edulis* roots and rhizomes with their retention indices (CP Sil 5 column, molecular weight (BP⁺) and base peak (bp; 100% of the mass spectrum)) (Bos *et al.* 1997b,c,d).

*V. officinalis**

Component	Retention Index	M^+	BP
Borneol	1141	154	95
Bornyl acetate	1262	196	43
Kessane	1505	222	43
Valerianol	1624	222	59
Valeranone	1639	222	41
Cryptofauranol	1644	238	41
Valerenal	1686	218	91

V. wallichii

Component	Retention Index	M^+	BP
Borneol	1141	154	95
Bornyl acetate	1264	196	43
Maaliol	1541	222	69
Patchouli alcohol	1625	222	83
Xanthorrizol	1717	218	136
a-Kessyl acetate	1772	280	43

V. edulis

Component	Retention Index	M^+	BP
Patchoulene	1423	204	122
Patchouli alcohol	1625	222	83

[1]Based on the principal components of the oil, four chemotypes can be distinguished within the species *V. officinalis*: the valeranone, valerianol, cryptofauranol and valerenal types (Bos *et al.*, 1986). Initially, the valerianol type was called elemol type (Hendriks *et al.*, 1977, Hazelhoff *et al.*, 1979a; Hendriks and Bruins 1980)

Gas chromatography (GC)

GC analysis was performed using WCOT fused-silica CP-Sil 5 CB, 25 m x 0.32 mm i.d.; film thickness, 0.25 μm (oven temperature programme, 50-290°C at 4°C/min; injection temperature, 250°C; detector (FID) temperature, 300°C; carrier gas, nitrogen; inlet pressure, 5 psi; linear gas velocity, 26 cm/s; split ratio, 1:56; injected volume, 1.0 μl.

Gas chromatography-mass spectrometry (GC-MS)

GC-MS (EI) was performed using the conditions described above, except: column, 25 m x 0.25 mm i.d.; carrier gas, helium; linear gas velocity, 32 cm/s; split ratio, 1:20. MS conditions: ionization energy, 70 eV; ion source temperature, 250°C; interface temperature, 280°C; scan speed, 2 scans/s; mass range, 34-500 amu; injected volume, 1.0 μl.

The identity of the components was assigned by comparison of their retention indices, relative to C_9-C_{19} n-alkanes, and mass spectra with corresponding data from reference compounds and from the literature (Adams 1989; Tucker and Maciarello 1993). The concentration of the components was calculated from the GC peak areas, using the normalization method.

In Table 2 the main constituents of the essential oils of *V. officinalis*, *V. wallichii* and *V. edulis* roots and rhizomes with their retention indices are listed.

Mixed Constituents

Thin-layer chromatograpy (TLC)

In our laboratory the following TLC system is applied in order to prove the identity of *V. officinalis*. An 1:5 70% ethanolic extract of plant material is made and chromatographed on silica gel using a mobile phase of hexane-diethyl ether 6:4. Detection uses anisaldehyde/sulphuric acid reagent followed by 5-10 min heating at 105°C.. The detected compounds with their respective R_f values and the colours of the spots are listed in Table 3.

High-performance liquid chromatography (HLC)

Recently, we developed a sensitive on-line HPLC method with diode array detection, by means of which valerenic acid and its derivatives, as well as valepotriates and baldrinals can be detected in one run (Bos *et al.,* 1996a). The procedure can be applied to both crude plant material and phytomedicines.

In the case of plant material H10.0 g is ground (1 mm) and extracted with 3 x 30 ml of methanol during 5 min in an ultrasonic bath (Bransonic 220) at room temperature. The extracts were filtered into a volumetric flask and the volume was adjusted to 100.0 ml with methanol. This extract was submitted to HPLC analysis as described below.

Table 3 Characteristic thin-layer chromatogram of the main components of *V. officinalis* (precoated channeled glass plates with 250 μm silica gel layer; hexane-diethyl ether (6:4); 15 cm; anisaldehyde sulphuric acid reagent)

Compound	R_f	Colour
Hydroxyvalerenic acid	0.01	Violet
Acetoxyvalerenic acid	0.15	Violet
Valerenic acid	0.38	Violet
Baldrinal	0.47	Yellow
Cryptofauronol	0.65	Purple-violet
Patchouli alcohol	0.70	Brownish-blue
Valerenal	0.86	Blue
Valeranone	0.89	Yellow

Table 4   Reference valerian compounds, separated using the on-line HPLC System (Bos et al., 1996a), with their respective retention time, capacity factor and UV maximum.

Compound	Retention time (min)	Capacity factor*	UV maximum (nm)
Baldrinal	5.00	1.99	424
Hydroxyvalerenic acid	5.39	2.21	220-221
Homobaldrinal	11.14	5.60	424
Acetoxyvalerenic acid	11.79	6.02	220-221
Acevaltrate	19.22	10.44	255
Valerenic acid	20.28	11.07	220-221
Didrovaltrate	20.31	11.09	200
Isovaleroxyhydroxy- didrovaltrate	21.16	11.60	200
Isovaltrate	22.14	12.18	255
Valtrate	22.92	12.64	255

*The capacity factor (k') was calculated using the formula $k' = T_R - T_0 / T_0$; T_R = retention time of peak (min) and T_0 = retention time of uracil (void time).

If phytomedicines such as tinctures, capsules and coated tablets, are analysed using the HPLC method some modifications are made in the preparation process. Tinctures are analysed as such, but the content of a capsule is first dissolved in 100.0 ml methanol and a coated tablet in 5.0 ml methanol. All samples were filtered through a 45 μm DynaGard HPLC filter (Microgon Inc., Laguna Hills, CA, USA) before injection into the HPLC apparatus.

HPLC was performed using an Isco HPLC pump 2350, an Isco gradient mixer 2360, a Kontron autosampler 360, an Isco V4 absorbance detector, a Kontron PC Integration pack and a Shimadzu SPDM6A-Diode Array Detector (DAD) under the following conditions. DAD: wavelength, 200-600 nm; band width, 2 nm; spectrum abs. scale (mAbs), 10-500; normalization threshold (mAbs), 10; analytical column: Superspher 100 RP-18 (5 μm) (LiChrocart 250-4); guard column: LiChrospher 100 RP-18 (5 μm) (LiChrocart 4-4) (Merck, Darmstadt, Germany); eluent A: 800 g water + 156.4 g acetonitrile; eluent B: 200 g water + 625.6 g acetonitrile (both eluents contain 1 mM phosphoric acid); flow rate: 1.5 ml/min; gradient: first 55% A and 45% B for 5 min, then up to 100% B in 19 min, followed by 100% B for 2 min, subsequently back to 45% B in 2 min, and finally again 55% A and 45% B for 5 min; start pressure: 22.5 MPa, decreasing to 14.5 MPa; injected volume: 20 μl.

In Table 4 reference valerian compounds, separated with the HPLC system, are listed with their respective retention time, capacity factor and UV maximum. Compounds with about the same retention times could be distinguished by their UV maximum.

STORAGE CONDITIONS

According to the European Pharmacopoeia 2nd edn (1993), the crude drug should be stored in a well-closed container, protected from light.

Powdered valerian root rapidly loses its essential oil and will no longer comply the pharmacopoeial standard. In addition, the drying as well as the storage temperature are important. At temperatures higher than 40°C, the valepotriates start to decompose, yielding valeric and isovaleric acids. The characteristic odour of these acids points to improperly dried or stored material. Additionally, hydroxyvalerenic acid may be considered to be a decomposition product of acetoxyvalerenic acid when the drug is stored at a too high humidity (Freytag 1983; Bos et al., 1996a)

Another important aspect is the stability. Valerenic acid and its derivatives have been proved to be stable in both plant material and preparations. In contrast, valepotriates rapidly decompose if water is present. They show a temperature-dependent instability. In commercially available valerian tinctures, for instance, valepotriates could not be detected anymore. After a storage time of several weeks these compounds have decomposed, first yielding baldrinals, that may react further with yet unknown substances present in the tincture or form polymerization products (Bos et al., 1996a).

REGULATORY ASPECTS

Present Situation

For quality control of the crude drug and of phytomedicines prepared from valerian, monographs of the European Pharmacopoeia 2nd edn (1993) can be used, but additional analyses are necessary (Woerdenbag 1995).

Between various countries of the European Union and in the United States of America, large differences exist in the legal position of herbal medicines. Regulations vary considerably from country to country, and in most European countries and in the USA the legislation of herbal drugs is much less progressive than for synthetic drugs.

The German health authorities, 'Bundesgesundheitsamt', have set up an expert committee for the evaluation of herbal drugs, the so-called 'Kommission E'. The findings of the Kommission E, based on data available from literature, are laid down in a monograph, in which the balance between usefulness and risk is weighed (Woerdenbag et al., 1993; Woerdenbag 1995). For valerian a 'Positiv-Monographie' exists, meaning that the herbal drug has been found to be biologically active, without inducing serious side-effects (Anonymous 1992). The root is permitted for oral use, and as herbal tea and tincture (De Smet 1993).

In France and Belgium, herbal remedies are subject to general drug regulations, and should therefore comply with criteria of efficacy, safety and quality. As compared with synthetic drugs, a simplified admission procedure is applied, which is based on chemical and pharmaceutical documentation, as available from the literature. Subterranean parts of V. officinalis and preparations are permitted for oral use in France. In Belgium subterranean parts, powder, extract and tincture are permitted as traditional tranquillizer (De Smet 1993).

In the United Kingdom, there are no special guidelines for the admission of herbal remedies to the drug market. To be accepted as a medicinal product, a herbal preparation must comply with the Guidelines on Safety and Efficacy Requirements for Herbal Medicinal Products. As proofs of safety and efficacy, literature data should be submitted.

For acceptance as an approved drug, however, evidence from clinical trials is required. For this reason, only few phytomedicines, not including valerian, have reached the status of approved drug in the United Kingdom (De Smet 1993).

In the Netherlands, no other legal regulations than the European Pharmacopoeia 2nd edn (1993) and the 'Warenwet' (Food Act) exist for phytomedicines. They do not have the status of drugs therefore, and may not be called a drug or recommended as such. In principle, herbal preparations can be sold by anyone in the Netherlands, a situation which tends to provoke inexpert use (Woerdenbag 1995).

It is clear that there is no unity at all in the European Union, in the field of herbal drugs. In 1989 the European Scientific Cooperative on Phytotherapy (ESCOP) was founded. The general aim of ESCOP is to advance the scientific status of phytomedicines and to assist with harmonization of their regulatory status in European countries. In a European frame-work, ESCOP prepares monographs for herbal drugs. These monographs, officially known as SPCs (Summary of Product Characteristics) are offered to regulatory authorities, the Committee on Proprietary Medicinal Products (CPMP), as a means of harmonizing the medicinal uses of plant drugs in the European Union and in a wider European context. An SPC of *Valerianae radix* has been submitted to the CPMP (Krant 1994).

ESCOP defines phytomedicines as follows:

> Phytomedicines (plant medicines) are medicinal products containing as active ingredient only plants, parts of plants or plant materials, or combinations thereof, whether in the crude or processed state.

Many of today's widely used herbs were once the subject of official monographs in The United States Pharmacopeia and The National Formulary, including valerian root (*V. officinalis*). No such legal standards exist in the USA today. The Food and Drug Administration (FDA) has evaluated the safety and efficacy of several herbal medicines. Valerian root has been granted the status 'Generally Recognized As Safe' (GRAS) (De Smet 1993; Tyler 1993, 1994).

Future Developments

In the near future the European Pharmacopoeia Commission is expected to change the monograph on valerian root. An HPLC assay for the quantitative determination of valerenic acid will be included, as well as a demand for its content in the crude drug. The present demand of at least 0.5% (v/m) essential oil will then probably be abandoned (J.H. Zwaving, personal communication). The quantitative determination of valerenic acid derivatives ensures that extracts are not adulterated by roots of *V. wallichii* and/or *V. edulis* (Hänsel and Schulz 1985).

Hänsel and Schulz (1985) proposed criteria for the content of valerenic acid and its derivatives. They recommended that alcoholic extracts of *V. officinalis* should contain valerenic acids with a minimum of 120 mg/100 g and aqueous extracts a minimum of 60 mg/100 g.

Valerian-containing phytomedicines are sometimes standardized, but this is not always the case (Bos *et al.,* 1996a). Preferably this should be done on valerenic acid and its derivatives, because of their stability. This implies that *V. officinalis* is the crude material of choice for such phytomedicines (Hänsel 1992).

For the European Pharmacopoeia a monograph on *Extractum Valerianae siccum* (dry valerian extract) is in preparation (Zwaving 1993).

Note

As of 1 January 1997, the 2nd edition of the European Pharmacopoeia has been replaced by the 3rd edition. The monograph 'Valerian root' (Valerianae radix) in the current edition has hardly changed as compared with the former. Therefore, in all cases that is referred to the 2nd edition of the European Pharmacopoeia, it can also be read as being referred to the 3rd edition.

European Pharmacopoeia, 3rd edition (1997), Council of Europe, Strasbourg.

REFERENCES

Adams, R.P. (1989) *Identification of Essential Oils by Ion Trap Mass Spectroscopy*, Academic Press, Inc., San Diego, New York, Berkeley, Boston, London, Sydney, Tokyo, Toronto.

Anonymous (1992) Bekanntmachung über die Zulassung und Registrierung von Arzneimitteln. *Bundesanzeiger*, No. **90**, May 15, 1988.

Böhme, H. and Hartke, K. (1978) Europäisches Arzneibuch. Band I und Band II. Kommentar. Wissenschaftliche Verlagsgesellschaft mbH, Stuttgart, Govi-Verlag GmbH, Frankfurt.

Böhme, H. and Hartke, K. (1979) Europäisches Arzneibuch. Band III. Kommentar. Wissenschaftliche Verlagsgesellschaft mbH, Stuttgart, Govi-Verlag GmbH, Frankfurt, p. 833–837.

Bos, R., Van Putten, F.M.S., Hendriks, H. and Mastenbroek, C. (1986) Variations in the essential oil content and composition in individual plants obtained after breeding experiments with a *Valeriana officinalis* strain. In E.-J. Brunke, (ed.), *Progress in Essential Oil Research*, Walter de Gruyter & Co., Berlin, p. 123–130.

Bos, R., Woerdenbag, H.J., Hendriks, H. and Malingré, Th.M. (1992) Der indische oder pakistanische Baldrian. *Z. Phytother.*, **13**, 26–34.

Bos, R., Woerdenbag, H.J. and Zwaving, J.H. (1994) Valeriaan en valeriaanpreparaten. *Pharm. Weekbl.*, **129**, 37–43.

Bos, R., Woerdenbag, H.J., Hendriks, H., Zwaving, J.H., De Smet, P.A.G.M., Tittel, G., Wikström, H.V. and Scheffer, J.J.C. (1996a) Analytical aspects of phytotherapeutic valerian preparations. *Phytochem. Anal.*, **7**, 143–151.

Bos, R., Woerdenbag, H.J., Hendriks, H., Sidik, Wikström, H.V. and Scheffer, J.J.C. (1996b) The essential oil and valepotriates from roots of *Valeriana javanica* Blume grown in Indonesia. *Flavour Fragr. J.*, **11**, 321–326.

Bos, R., Woerdenbag, H.J., De Smet, P.A.G.M. and Scheffer, J.J.C. (1997a) *Valeriana* species. In P.A.G.M. De Smet, K. Keller, R. Hänsel, and R.F. Chandler, (eds.), *Adverse Effects of Herbal Drugs, Volume 3*, Springer-Verlag, Berlin, Heidelberg, p. 165–180.

Bos, R., Woerdenbag, H.J., Hendriks, H., Smit, H.F., Wikström, H.V. and Scheffer, J.J.C. (1997b) Composition of the essential oil from roots and rhizomes of *Valeriana wallichii* DC. *Flavour Fragr. J.*, 123–131.

Bos, R., Woerdenbag, H.J., Hendriks, H. and Scheffer, J.J.C. (1997c) Composition of the essential oils from underground parts of *Valeriana officinalis* L. s.l., and several closely related taxa.

Bos, R., Woerdenbag, H.J., Hendriks, H. and Scheffer, J.J.C. (1997d) Essential oil from roots of Mexican *Valeriana* species, in preparation.

Bounthanth, C., Bergmann, C., Beck, J.P., Haag-Berrurier, M. and Anton, R. (1981) Valepotriates, a new class of cytotoxic and antitumor agents. *Planta Med*, **41**, 21–28.

Bounthanth, C., Richert, L., Beck, J.P., Haag-Berrurier, M. and Anton, R. (1983) The action of valepotriates on the synthesis of DNA and proteins of cultured hepatoma cells. *Planta Med*, **49**, 138–142.

Braun, R., Dittmar, W., Machut, M. and Wendland, S. (1983) Valepotriate - zur Bestimmung mit Hilfe von Nitrobenzylpyridin (NBP-Methode). *Dtsch. Apoth. Ztg.*, **123**, 2474–2477.

Braun, R., Dieckmann, H., Machut, M., Echarti, C. and Maurer, H.R. (1986) Studies on the effects of baldrinal on hemopoietic cells *in vitro*, on the metabolic activity of the liver *in vivo*, and on the content in proprietry drugs. *Planta Med.*, **52**, 446–450.

Chavadej, S., Becker, H. and Weberling, F. (1985) Further investigations of valepotriates in the Valerianaceae. *Pharm. Weekbl. Sci. Ed.*, **7**, 167–168.

De Smet, P.A.G.M. and Vulto, A.G. (1988) Drugs used in non-orthodox medicine. In M.N.G. Dukes and L. Beely, (eds.), *Side Effects of Drugs - Annual 12*, Elsevier, Amsterdam, p. 402–415.

De Smet, P.A.G.M. (1993) Legislatory outlook on the safety of herbal remedies. In P.A.G.M. De Smet, K. Keller, R. Hänsel and R.F. Chandler, (eds.) *Adverse Effects of Herbal Drugs, Volume 2,* Springer-verlag, Berlin, p. 1–90.

Dieckmann, H. (1988) Untersuchungen zur Pharmakokinetik, Metabolismus und Toxikologie von Baldrinalen. Freie Universität Berlin (dissertation).

Dossaji, S.F. and Becker, H. (1981) HPLC separation and quantitative determination of valepotriates from *Valeriana kilimandscharica*. *Planta Med.*, **43**, 179-182.

Dutch Pharmacopoeia, '*Nederlandse Farmacopee*', 6th edn, 2nd printing (1966), Staatsuitgeverij, 's-Gravenhage, p. 72.

European Pharmacopoeia, 1st edn (1975), Council of Europe, Strasbourg.

European Pharmacopoeia, 2nd edn (1993), Council of Europe, Strasbourg.

Evans W.C. (1989) *Trease and Evans' Pharmacognosy*, 13th edn., Baillière Tindall, London, p. 525–528.

Förster, W., Becker, H. and Rodriguez, E. (1984) HPLC analysis of valepotriates in the North American genera *Plectritis* and *Valeriana*. *Planta Med.*, **50**, 7–9.

Freytag, W.E. (1983) Bestimmung von Valerensäuren und Valerenal neben Valepotriaten in *Valeriana officinalis* durch HPLC. *Pharm. Ztg.*, **128**, 2869–2871.

German Homoeiopathic Pharmacopoeia, '*Homöopathisches Arzneibuch*', 1. Ausgabe (1975), and 3. Nachtrag (1985), Deutscher Apotheker Verlag, Stuttgart; Govi-Verlag GmbH, Frankfurt, p. 415–418.

German Pharmacopoeia, '*Deutsches Arzneibuch*', 10. Auflage (1993), Deutscher Apotheker Verlag, Stuttgart; Govi-Verlag GmbH, Frankfurt.

Graf, E. and Bornkessel, B. (1978) Analytische und pharmazeutisch-technologische Versuche mit Baldrian. *Dtsch. Apoth. Ztg.*, **118**, 503–508.

Gränicher, F., Christen, P. and Kapetanidis, I. (1992) High-yield production of valepotriates by hairy root cultures of *Valeriana officinalis* L. var. *sambucifolia* Mikan. *Plant Cell Rep.*, 11, 339–342.

Hänsel, R. and Schulz, J. (1982) Valerensäure und Valerenal als Leitstoffe des offizinellen Baldrians. *Dtsch. Apoth. Ztg.*, **122**, 215–219.

Hänsel, R., Schulz, J. and Stahl, E. (1983) Prüfung der Baldrian-Tinktur auf Identität. *Arch. Pharm.* (Weinheim), **316**, 646–647.

Hänsel, R. and Schulz, J. (1985) Beitrag zur Qualitätssicherung von Baldrianextrakten. 4. Mitt. *Pharm. Ind.*, **47**, 531–533.

Hänsel, R. (1990) Pflanzliche Sedativa. *Z. Phytother.*, **11**, 14–19.

Hänsel, R. (1992) Indischer Baldrian nicht empfehlenswert? *Z. Phytother.*, **15**, 130–131.

Hartke, K and Mutschler, E. (1987) DAB 9 - Kommentar. Band 2. Wissenschaftliche Verlagsgesellschaft mbH, Stuttgart, Govi-Verlag GmbH, Frankfurt, p. 917–921.

Hazelhoff, B., Smith, D., Malingré, Th.M. and Hendriks, H. (1979a) The essential oil of *Valeriana officinalis* L. s.l. *Pharm. Weekbl.*, **114**, 443–449.

Hazelhoff, B., Weert, B., Denee, R. and Malingré, Th.M. (1979b) Isolation and analytical aspects of Valeriana compounds. *Pharm. Weekbl. Sci. Ed.*, **1**, 140–148.

Hendriks, H. and Bruins, A.P. (1980) Study of three types of essential oil of *Valeriana officinalis* L. s.l. by combined gas chromatography-negative ion chemical ionization mass spectrometry. *J. Chromatogr.*, **190**, 321-330.

Hendriks, H., Smith, D. and Hazelhoff, B. (1977) Eugenyl isovalerate and isoeugenyl isovalerate in the essential oil of Valerian root. *Phytochemistry*, **16**, 1853–1854.

Hendriks, H. and Bos, R. (1984) Essential oils of some Valerianaceae. *Dragoco Rep.* (English ed.) **1**, 3–17.

Houghton, P. (1988) The biological activity of valerian and related plants. *J. Ethnopharmacol.*, **22**, 121–142.

Indian Pharmacopoeia, 2nd edn. (1966) Manager of Publications, New Delhi.

Keochanthala-Bounthanth, C., Haag-Berrurier, M., Beck, J.P. and Anton, R. (1990) Effects of thiol compounds versus cytotoxicity of valepotriates on cultured hepatoma cells. *Planta Med.*, **56**, 190–192.

Keochanthala-Bounthanth, C., Beck, J.P., Haag-Berrurier, M. and Anton, R. (1993) Effects of two monoterpene esters, valtrate and didrovaltrate, isolated from *Valeriana wallichii*, on the ultrastructure of hepatoma cells in culture. *Phytother. Res.*, **7**, 124–127.

Krant, W. (1994) Milestones for ESCOP. *Europ. Phytotelegram*, **6**, 4–6.

Laufer, J.L., Seckel, B.J. and Zwaving, J.H. (1970) Onderzoek naar de samenstelling van de werkzame bestanddelen van verschillende valeriaan- en kentranthussoorten. *Pharm. Weekbl.* **105**, 609–625.

Lipták, J. and Verzár-Petri, G. (1980) Titrimetrische Bestimmung der Valepotriate. *Sci. Pharm.*, **48**, 203–206.

Lorens, M. (1989) Untersuchungen zur Domestikation der mexikanischen Medizinalpflanze *Valeriana edulis* ssp. *procera* Meyer. Technischen Universität München, Freising-Weihenstephan, Lehrstuhl für Gemüsebau (dissertation).

Mannestätter, E., Gerlach, H and Poethke, W. (1967) Über die Inhaltsstoffe von Valerianaceen. 1. Mitt.: Der Nachweis einiger Inhaltsstoffe von *Kentranthus ruber* DC. *Pharm. Zentralhalle*, **106**, 797–804.

Mannestätter, E., Gerlach, H and Poethke, W. (1968) Über die Inhaltsstoffe von Valerianaceen. 3. Mitt.: Beiträge zur Bestimmung des Halazuchroms B in Valerianaceen-Drogen. *Pharm. Zentralhalle*, **107**, 261–269.

Morazzoni, P. and Bombardelli, E. (1995) *Valeriana officinalis*: traditional use and recent evaluation of activity. *Fitoterapia*, **66**, 99–112.

Rücker, G., Neugebauer M. and Sharaf El Din, M. (1981) Quantitative Dünnschichtchromatographische Analyse von Valepotriaten. *Planta Med.*, **43**, 299–301.

Schimmer, O. and Röder, A. (1992) Valerensäuren in Fertigarzneimitteln und selbst bereiteten Auszügen aus der Wurzel von *Valeriana officinalis* L. s.l. *Pharm. Ztg. Wiss.*, **137**, 31–36.

Stahl, E. and Schild, W. (1969) Dünnschicht-Chromatographie zur Kennzeichnung von Arzneibuchdrogen. *Arzneim.-Forsch.*, **19**, 314–316.

Stahl, E. and Schütz, E. (1980) Extraktion labiler Naturstoffe mit überkritischen Gasen. *Planta Med.*, **40**, 262–270.

Stahl, E. (1970) Chromatographische und mikroskopische Analyse von Drogen. Fischer-Verlag, Stuttgart, p. 161.

Steinegger, E. and Hänsel, R. (1992) Pharmakognosie, 5th edn. Springer-Verlag, Berlin, p. 162-163, 666-671.

Thies, P.W. (1968) Die Konstitution der Valepotriate. Mitteilung über die Wirkstoffe des Baldrians. *Tetrahedron*, **24**, 313–347.

Thies, P.W. (1969) Zum chromogenen Verhalten der Valepotriate. *Arzneim.-Forsch.*, **19**, 319-322.

Thies, P.W. and Funke, S. (1966) Nachweis und Isolierung von sedativ wirksamen Iso-Valeriansäureestern aus Wurzel und Rhizomen von verschiedenen *Valeriana*- und *Kentranthus*-Arten. *Tetrahedron Lett.*, **11**, 1155–1162.

Tittel, G. and Wagner, H. (1978) Hochleistungsflüssigchromatographische Trennung und quantitative Bestimmung von Valepotriaten aus Valeriana-Drogen und Zubereitungen. *J. Chromatogr.*, **148**, 459–468.

Tittel, G., Chari, V.M. and Wagner, H. (1978) HPLC-Analyse von *Valeriana mexicana* Extrakten. *Planta Med.*, **34**, 305–310.

Titz, W., Jurenitsch, J., Fitzbauer-Busch, E., Wicho, E. and Kubelka, W. (1982) Valepotriate und ätherisches Öl, morphologisch und chromosomal definierter Typen von *Valeriana officinalis* L. s.l. I. Vergleich von Valepotriatgehalt und -Zusammensetzung. *Sci. Pharm.*, **50**, 309–324.

Titz, W., Jurenitsch, J., Gruber, J., Schabus, L., Titz, E. and Kubelka, W. (1983) Valepotriate und ätherisches Öl, morphologisch und chromosomal definierter Typen von *Valeriana officinalis* L. s.l. II. Variation charakteristischer Komponenten des ätherischen Öls. *Sci. Pharm.*, **51**, 63–86.

Tucker, A.O. and Maciarello, M. (eds.) (1993) *Mass Spectral Library of Flavor & Fragrance Compounds, Volume I-XXVII*, Department of Agriculture & Natural Resources, Delaware State University, Dover, Delaware.

Tyler, V.E. (1993) *The Honest Herbal*, 3rd edn, Pharmaceutical Products Press, New York, p. 315–317, p. 350–351.

Tyler, V.E. (1994) *Herbs of Choice*, Pharmaceutical Products Press, New York, p. 17-32, 117–119.

Van Meer, J.H., Van der Sluis, W.G. and Labadie, R.P. (1977) Onderzoek naar de aanwezigheid van valepotriaten in valeriaanpreparaten. *Pharm. Weekbl.*, **112**, 20–27.

Van Meer, J.H. and Labadie, R.P. (1981) Straight-phase and reversed-phase high-performance liquid chromatographic separations of valepotriate isomers and homologues. *J. Chromatogr.*, **205**, 206–212.

Verzár-Petri, G., Pethis, E. and Lemberkovics, E. (1976) Iridoid jellegü vegyületek réteg - és gázkromatográfiás vizsgálata a *Valeriana officinalis*. L. - ben. *Herba Hung.*, **15**, 79–91.

Von der Hude, W., Scheutwinkel-Reich, M., Braun, R. and Dittmar, W. (1985) *In vitro* mutagenicity of valepotriates. *Arch. Toxicol.*, **56**, 267–271.

Von der Hude, W., Scheutwinkel-Reich, M. and Braun, R. (1986) Bacterial mutagenicity of the tranquilizing constituents of Valerianaceae roots. *Mutat. Res.*, **169**, 23–27.

Wagner, H., Hörhammer L., Höltzl, J. and Schaette, R. (1970) Zur Wertbestimmung der Baldrian-Droge und ihrer Zubereitungen. *Arzneim.-Forsch.*, **20**, 1149–1152.

Wichtl, M. (1989) *Teedrogen*, 2nd edn, Wissenschaftliche Verlagsgesellschaft mbH, Stuttgart, p. 79–82.

Wienschierz, H.J. (1978) Erfahrungen bei der Kultivierung von *Valeriana wallichii* (DC) in der Bundesrepublik Deutschland. *Acta Horticult.*, **73**, 315–321.

Woerdenbag, H.J., De Smet, P.A.G.M. and Scheffer, J.J.C. (1993) Plantaardige geneesmiddelen in medisch-farmaceutisch perspectief. *Pharm. Weekbl.*, **128**, 164–177.

Woerdenbag, H.J. (1995) The role and research of phytomedicines in European countries. *Ned. Tijdschr. Fytother.*, **8**, (1), 13–16.

Zwaving, J.H. (1993) Kwaliteitscontrole van natuurlijke grondstoffen door de apotheker en zijn rol in de voorlichting over plantaardige geneesmiddelen aan de patiënt. *Pharm. Weekbl.*, **128**, 178–184.

6. *VALERIANA* PRODUCTS

RICHARD FOSS[1] and PETER J HOUGHTON[2]

[1]*Agros Associates, Yew Tree House, School House Lane, Aylsham, Norfolk NR11 6EX, U.K.*
[2]*Pharmacognosy Research Laboratories, Department of Pharmacy, King's College London, Manresa Road, London SW3 6LX*

CONTENTS

INTRODUCTION

There are four species of *Valeriana* that are important articles of commerce either as the plant material or as extracts used in the production of the commodities mentioned below. These species are European Valerian *V. officinalis* L., Indian Valerian *V. wallichii* DC., Mexican Valerian *V. edulis* Nutt. ex Torr. & Gray and Japanese Valerian *V. fauriei* Briq.. Commercial supplies of these four species are mainly obtained from cultivation but some plants are still collected from the wild. The first three of these are cultivated in Europe whilst Japanese Valerian is grown and used mainly in the Far East and Indian Valerian is the species grown and used on the Indian subcontinent. *V. officinalis* is also grown commercially in North America. Most of the data available refers to *V. officinalis* since this is the species which has received most attention as a commercial crop and is consequently utilised in Western society. It should be remembered, however, that a large trade in these and more local *Valeriana* species, as with other plants used in traditional

Table 1 Markets for products containing Valeriana officinalis

Country	Number of products	Number of products containing raw materials					Main manufacturers	Major therapeutic uses	Valerian as sole active ingredient	
		Base	Extract	Oil	Powder	Tincture			Number	Examples
Argentina	4	2	1			1	Nora	Sedative	0	
Australia	10	6	4				Rhone Poulenc, Vitaplex	Sedative	5	Polcopharma Valerian, B/G Valerian, N/W Valerian, Soul Pattinson Valerian, Vitaglow Valerian
Austria	30	13	16			1	Smetana, Chepharin-Hauser, Lyssia	Sedative	6	Baldrian Kneipp, Baldrian-Drei-Herzbl, Valin Baldrian, Baldrian Klosterfrau, Baldrian Dispert, Baldrinetten
Belgium	7	3	3				Centrapharm, Lyssia	Sedative	0	Valdispert
Brazil	5	2	3				Leofarm	Sedative	0	
Canada	4	3				1	MRF	Sedative	2	Valerian
Central America	5	3	1	1			Vida, Henmann	Sedative	0	
Chile	4		4				Medipharm, Volta	Sedative, anti-spasmodic	3	Nerviol, Tintura de Valeria
Colombia	2	2					Inquifar, Lafont	Analgesis, antitussive	0	
Egypt	2	1	1				Mepaco	Sedative	0	
Finland	2	1	1				Lerius, Orion	Sedative	2	Valrian, Valdispert
France	57	24	27		3	1	Ardeval, Lehning, Monol, Arkopharma	Sedative	5	Valeriane Effidose, Valerian Vitaflor, Valeriane Pachant, Valeriane Titrex, Valeriane Arkogelules

Table 1 continued

Country	Number of products	Number of products containing raw materials					Main manufacturers	Major therapeutic uses	Valerian as sole active ingredient	
		Base	Extract	Oil	Powder	Tincture			Number	Examples
Franco-phone West Africa	14	4	7		1	2	Lehning	Sedative	0	
Germany	471	193	195	18	2	58	Fides, Hanosan, Nestmann, Pascoe, Pharmakon, Phoenix, Schuck, Spitzner, Weber und Weber	Sedative, cardiovascular	34	Recvalysat, Abtei Baldrian, Baldrianwurzel, Baldriafur, Sedalint Baldrian, Baldrain Dispert, Valdispert, Kohl Baldrian Dispert, Baldrian Drag Lip, Baldrian-Phyton, Baldrisedon, Orasedon, Baldrianetten, Visinal Beruhigung, Regivital Baldrian, Togasan Baldrian, Balsedat, Knufinke Baldrian, Baltherm Baldrian, Perozon Baldrian, Baldrian Tinktur, Functional Baldrian
Greece	2		2				Chrop 1, Santa	Sedative	0	
Hungary	2		2				Biogal	Sedative	0	
Italy	23	2	20			1	Edmond Pharma, SIT	Sedative	5	Val uno, Valeriana Farmades, Valeriana Dispert, Tintura Valeriana

Table 1 continued

Country	Number of products	Number of products containing raw materials					Main manufacturers	Major therapeutic uses	Valerian as sole active ingredient	
		Base	Extract	Oil	Powder	Tincture			Number	Examples
Japan	13	1	7		5		Zaiseido	Sedative, antihypertensive, cardiovascular	0	
Korea	10	5	3		2		Han Lim, Hae Woi	Cardiovascular	0	
Mexico	5		4				Franco Mexicana	Sedative	0	
Morocco	1		1			1	Soekami-Lefranq	Sedative	0	
Netherlands	15	2	12	1			Pharbita, Daro, Roche Nicholas, VSM	Sedative	9	Daro Valerian, Extract Valerianae, Valeriaan Katwyk, Calmolan, Tendo Valeriaan, Valdispert, Extract Valeriane-2
New Zealand	7	6				1	Blackmores, Red Seal	Sedative	3	Blackmores Valerian, Meadow Croft Valerian, Valerian Compound
Pakistan	2	2					Rafi, Uni Herbal	Gynaecological, antacid	0	
Philippines	1		1				Arnet Pharmaceutical		0	
Portugal	11	5	6				Zyma Farma Portugeas, Roba Portugesa	Sedative	2	Valdispert, Circulin
South Africa	5	2	1		1	1	Noristan	Sedative	1	Calmettes
Spain	9	6	2		1	1	Deiters, Ordesa	Sedative	6	Valeriana Deiters, Valeriana Kneipp, Fitokey Valeriana, Relaxul Valeriana, Valeriana Nutter

Table 1 continued

| Country | Number of products | Number of products containing raw materials | | | | | Main manufacturers | Major thera-peutic uses | Valerian as sole active ingredient | |
		Base	Extract	Oil	Powder	Tincture			Number	Examples
Sweden	3	1	2				Kabi Pharmacia	Sedative	3	Valerecen, Neurol, Baldrian Dispert
Switzerland	86	33	39	5	1	8	Kuenzle, Dronamia, Robins, Wiedenmamm, Zeller	Sedative	5	Valdispert, Baldrisedon, Regivital, Kuenzle, Arkogelules
Taiwan	4	2	2				Sa Saoka Yakuhin	Sedative, gynaecological	0	
Tunisia	1				1		Sanofi Winthrop	Antitussive	0	
USA	2	2					Pharmavite	Sedative	2	Valerian root
UK	25	15	10				Gerard House	Sedative	2	
Venezuela	6	1	3		1	2	Lariviere, Gache	Sedative	1	Valerianato Pierl

medicine, occurs within developing countries at a local level and information concerning this usage is practically impossible to obtain. In these conditions material is more likely to be obtained from wild plants or from small-scale cultivation. The collection of wild material has raised concern about the possible extinction of less common species.

Valerian products are marketed and used worldwide (see Table 1) and comprise one of the best-selling entities of the health food/natural medicine sector in the industrialised world (Brevoort, 1996). More than two hundred commercial preparations containing *V. officinalis* are listed (Martindale, 1996). This chapter presents an overview of the market in the mid-1990s and the products available.

TRADE IN VALERIANA SPECIES

Production

About 1200 tonnes of *V. officinalis* roots are estimated to be produced annually in Europe in the years after 1990. Traditional growing areas include Germany, the Netherlands and Belgium and these still produce the best quality material. Recently large amounts of this species have been grown in eastern Europe, particularly in Poland, Bulgaria and Ukraine, and this has resulted in a surplus of an estimated 800 tonnes. *V. officinalis* is also produced for local consumption in North America (Hobbs, 1989).

V. wallichii originates from the Indian subcontinent and is still grown and used extensively there but is also now cultivated in China and Germany as a source of valepotriates (Bos *et al.,* 1992). *V. edulis* is collected and grown in Mexico, its country of origin, but is also cultivated in Germany, mainly as a source of valepotriates, since it contains high amounts of these compounds and has large roots.

The steam-distilled oil from *V. officinalis* is now produced mainly in China and the annual production is estimated at about 7 tonnes.

Markets

The major market for *Valeriana* at present in Europe is Germany where the retail sales value of valerian products is reckoned to be five million US dollars annually. Significant amounts of *Valeriana* products are also consumed in Switzerland, France, Austria and Italy.

Prices

There are two levels of prices for *V. officinalis*, the price at which the roots are sold by the growers and that at which they are sold by traders on the commodities market. In 1995 the first price was 2.00-2.50DM (US$ 1.4-1.7) per kg. Trader's selling prices per kilo varied according to the country where purchased and ranged from 6.5DM (US$4.4) in UK to 6.75-7.35DM (US$ 4.6-5.0) in France and 7.5-8.0DM (US$ 5.1-5.5) in Germany. It is expected that prices may be almost halved during the next few years because of the surplus of roots available.

The price of the volatile oil obtained from *Valeriana officinalis* has increased in recent years as cheap European products have been displaced by more expensive material of Chinese origin. In 1984 the price was US$42 per kilo but in 1993 was quoted at US$150 per kilo.

COMMERCIAL *VALERIANA* PRODUCTS

Medicinal products

In former days galenical preparations containing *V. officinalis* were much used in extemporaneous dispensing by pharmacists. The majority of medicinal preparations containing *Valeriana* products in some form are now sold in pharmacies or 'healthfood' retail outlets as mild sedatives and sleep inducers. The Table emphasises the worldwide use of such products.

In some countries, e.g. Germany, the dried root is still sold as a consumer product for the production of a tea to relieve over-excitement, aid the onset of sleep and for gastrointestinal disturbances (Bradley, 1992). Careful instructions for correct use and dosage have to be provided in some countries.

Extracts

The most common galenical preparation, listed in many official formularies and pharmacopoeias, is a tincture made with 60% ethanol. This may be concentrated to form a soft extract. However in some modern monographs a tincture made with 70% ethanol is described (DAB10, 1991). A concentrated infusion, made with 25% ethanol, has been employed in the UK (BPC, 1963). It is likely that the preparations made with more concentrated alcohol, when freshly prepared, contain some of the volatile oil components and also valepotriates thought to be the major active components, although the latter hydrolyse quickly upon storage.

Many commercial preparations contain dried extracts made from *Valeriana* species although the solvent used is not always specified.

Oils

The steam-distilled volatile oil (sometimes called the essential or ethereal oil) from *V. officinalis* is included in a few preparations (see Table) but it should be noted that it plays a much less prominent role than the various solvent extracts employed.

Commercial preparations

Valerian is included in commercial preparations either as powdered plant material or as a dry extract since these forms are the most easy to incorporate in oral dosage forms such as tablets and capsules. The Table gives an indication of the large number of commercial products which are available, many of which are marketed as 'over the counter' or self-selection products. In many countries there is little legal restriction at present on the sale of such entities since they are not officially classed as medicines.

It should be noted that only a minority of the products contain solely *Valeriana officinalis* as the active ingredient. In most preparations the powdered plant material or the extract is combined with those obtained from other plants which have a reputation for inducing relaxation or sleep. The most common of these are Passionflower Herb (*Passiflora incarnata* L.), Hops (the fruit of *Humulus lupulus* L.) and Scullcap Herb (*Scutellaria laterifolia* L.).

The dose of individual herbs in such preparations is not always as high as the that recommended in semi-official publications such as the British Herbal Pharmacopeia (1996) or German Kommission E monograph (1990, 1996) and careful selection and use of these products is advisable.

Isolated constituents

Although the majority of preparations based on *Valeriana* employ extracts or oils, there are a few instances where single compounds isolated from these crude extracts are used or where isolated compounds are combined in a standardised mixture.

The major example of the latter is Valmane ® which is a mixture of three valepotriates consisting of didrovaltrate 80%, valtrate 15% and acevaltrate 5%. It is used in several sedative proprietary preparations widely sold in European countries.

The single isolated constituents used are all obtained from the volatile oil distilled from the plant material. Valeric acid and isovaleric acid are used as active ingredients in some sedative products in Austria, Germany and Italy. The monoterpenes pinene and borneol are obtained from many other sources as well as *Valeriana*. They are used extensively in perfumery as well as in pharmaceutical products to relieve nasal congestion and for topical application as counter-irritants in the treatment of muscular pain. Pinene and borneol are also used as feedstock substances for the semisynthesis of a wide range of industrial and pharmaceutical fine chemicals.

Other uses

The essential oil of *Valeriana officinalis* is used to some considerable extent in the perfumery industry although it is blended in small amounts with other oils because of its strong odour which many find objectionable.

REFERENCES

Bos, R., Woerdenbag, H.J., Hendriks, H. and Malingré, Th. M. (1992) Der Indische oder Pakistanische Baldrian. *Zeitschrifte für Phytotherapie* **13**:26-34.
Brevoort, P. (1996) The U.S. botanical market - an overview. *Herbalgram* **36**: 49-57.
BPC (British Pharmaceutical Codex) 1963, Pharmaceutical Press, London p.1092.
Bradley, P.R. (editor) (1992) *British Herbal Compendium*, British Herbal Medicine Association, Bournemouth, UK p. 215-217.
British Herbal Pharmacopeia 1996, British Herbal Medicine Association, Bournemouth, UK p.176.

DAB 10 (1991) Deutsches Arzneibuch (German Pharmacopeia) 10th edition. Deutscher Apotheker Verlag, Stuttgart, Germany.

Hobbs, C. (1989) Valerian *Herbalgram* **21:**19-34.

Kommission E Monograph - Valerianae radix (Baldrianwurzel) - amendment (1990), Federal German Health Ministry, Bonn 13-3-90 (English translation - German Commission E Monographs (1996) American Botanical Council, Austin, Texas, USA)

Martindale - The Extra Pharmacopeia 31st edn(1996) ed. J.E.F. Reynolds, Pharmaceutical Press, London p 1766.

INDEX